W9-BCK-888

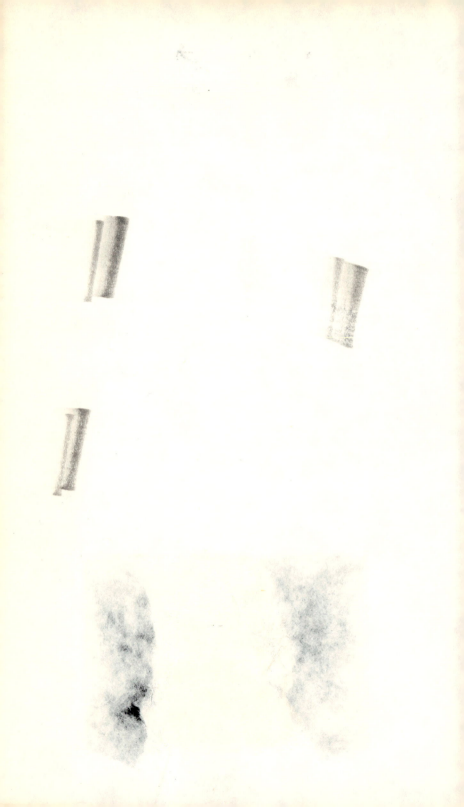

ICE AGES
RECENT AND ANCIENT

Freshfield Glacier. Canadian Rockies.

Photographed by A. O. Wheeler.

ICE AGES
RECENT AND ANCIENT

BY

A. P. COLEMAN

AMS PRESS
NEW YORK

Reprinted from the edition of 1926, New York
First AMS EDITION published 1969
Manufactured in the United States of America

QE
697
.C65
1969

Library of Congress Catalog Card Number: 77-105678
SBN: 404-01596-4

AMS PRESS, INC.
New York, N.Y. 10003

CONTENTS

INTRODUCTION

PART I

THE PLEISTOCENE ICE AGE

CHAPTER I

PLEISTOCENE GLACIATION IN NORTH AMERICA

CHAPTER II

THE ICE AGE IN OTHER REGIONS

CHAPTER III

RESULTS OF PLEISTOCENE GLACIATION

PART II

PRE-PLEISTOCENE ICE AGES

CHAPTER IV

CENOZOIC AND MESOZOIC GLACIATION

CHAPTER V

LATE PALÆOZOIC GLACIATION

CHAPTER VI

PERMO-CARBONIFEROUS GLACIATION IN INDIA

CHAPTER VII

PERMO-CARBONIFEROUS GLACIATION IN AFRICA

CHAPTER VIII

LATE PALÆOZOIC GLACIATION IN AUSTRALIA

CHAPTER IX

PERMO-CARBONIFEROUS GLACIATION IN SOUTH AMERICA

CONTENTS

CHAPTER X

PERMIAN OR CARBONIFEROUS GLACIATION IN NORTH AMERICA

CHAPTER XI

MIDDLE AND EARLY PALÆOZOIC GLACIATION

CHAPTER XII

GLACIATION AT OR JUST BEFORE THE BEGINNING OF THE CAMBRIAN

CHAPTER XIII

GLACIATION IN HURONIAN AND EARLIER TIMES

PART III

CAUSES OF GLACIATION

INTRODUCTION

CHAPTER XIV

GEOLOGICAL THEORIES OF ICE AGES

CHAPTER XV

ATMOSPHERIC AND ASTRONOMIC THEORIES

ILLUSTRATIONS

xi

MAPS

INTRODUCTION

Variations in Climate

SPACE is intensely cold so far as our knowledge goes. It is generally supposed to be at the absolute zero, 273° below the freezing point of water by the Centigrade thermometer. That is to say a body in space would have no temperature at all unless warmed by some internal or external source of heat. The motions of the molecules which show themselves as heat would be absent.

There are innumerable bodies scattered through space which we know to be intensely hot, since they shine brilliantly with their own light; but there are also innumerable dark bodies in space of which we know little, since they send us no messages. Some of them perhaps are devoid of heat—at the absolute zero.

Our earth is such a dark body, visible only when it reflects light coming from without, but it has a large store of heat in its interior. This has very little effect on the warmth of its surface, however, so that our temperatures and climates depend almost entirely on radiations coming from the sun, as modified and conserved by the atmosphere. We live in a glass house warmed by an infinitesimal part of the energy sent out by a furnace 92,000,000 miles away.

All life of plant or animal depends on the temperature of the surface of the earth and its envelopes. If we received a few degrees less of heat from the sun much of the world would be uninhabitable, and the same is true if our atmosphere became a less efficient blanket against loss of heat by radiation into space. On the other hand an important increase in the sun's radiation or of the power of the atmosphere to retain it might blast all life by the rise in temperature. Comparatively slight changes in the supply of heat would disarrange our whole economy if they did not destroy life. It is evident that the question of the permanence or variability of climates is a vital one.

Many articles have been written on the subject of variations in climate, with arguments for or against such changes within historic times. Professor Gregory quotes the date palm and the vine as sensitive thermometers, proving that the Mediterranean climates have not varied from the times of the Old Testament; while Professor Douglas measures the width of the annual rings of Californian sequoias, running back for 3,000 years, and finds cyclic changes of a somewhat important kind.

Alpine glaciers are shown to have advanced and retreated in a cyclic way within the last few hundred years; though, on the whole, glaciers all over the world seem to be retreating, suggesting increasing warmth or decreasing snowfall as the years go by. The once populous and wealthy cities now buried in the sands of the deserts of central Asia or the Sahara indicate cooler and moister conditions some thousands of years ago, and the peat deposits of Scandinavia show changes of the plant growth indicating warmer and drier or cooler and moister conditions within the last few thousand years. On the whole, however, such researches have not disclosed profound climatic changes since man built cities or made use of earthen tablets or of papyrus to record events. The changes seem to have kept within moderate limits since civilized man dwelt in Europe, Asia, or Northern Africa.

The results of the study of the last few thousand years

seem on the whole reassuring to the conservative mind, satisfied with things as they are.

It is not so, however, when one passes from human history into the more remote past, as shown by geology, where changes of climate of a most startling kind are disclosed, sometimes implying terrible catastrophes which half depopulated the globe, and at others suggesting mild or even tropical conditions in the polar regions.

During the meeting of the Geological Congress in Stockholm in 1910 the Swedish geologists arranged an excursion to Spitzbergen, and I well remember the delight with which we collected great fossil leaves belonging to trees of a warm climate in beds of Cenozoic age on the top of Mt. Nordenskjöld, with the snows of a glacier close by. The largest tree now growing in Spitzbergen is a sprawling willow, scarcely rising three inches above the ground and with leaves half an inch long.

A seam of excellent coal somewhat lower in the same formation shows how luxuriant the vegetation must have been. Rich forests like those of the middle United States then grew within twelve degrees of the pole.

More recently specimens of tree trunks as much as eighteen inches in diameter have been found along with coal even nearer to the South Pole in Antarctica. It is evident that the climate at certain times was mild from pole to pole.

On the other hand we have unmistakable evidence that near the end of the Palæozoic great ice sheets existed on low ground within the tropics on two continents and not far from the tropics on two others.

The Arctic regions have enjoyed genial periods and some parts of the Tropics have suffered Arctic cold, showing that in the past there have been very important variations of climate.

We are apt to think of our present conditions as normal and to look upon such wide swings of the climatic pendulum as those just mentioned as extremes; but there is good reason to believe that our present epoch is not really normal, but decidedly colder than usual, though much milder than in past ice ages. We are probably still in the closing stages of the Pleistocene Ice Age; the earth not yet warmed up to the usual level, as shown by the luxuriant life inhabiting the seas and lands of what are now temperate or even arctic regions during most of geological history.

In general the fossils preserved in the rocks do not show much evidence of zones of climate, and until recently geologists believed that during its earlier history the world had a hot-house character with little variation of temperature in the different latitudes. The evidence from ancient life was held to support the conclusion reached by certain philosophers and astronomers that the world was once intensely hot and has slowly been cooling down ever since. The nebular hypothesis was almost universally accepted by geologists till about a generation ago, and it was natural to think of the earth in the far past as without zones of climate and mild to the poles because its internal heat warmed up the surface and evaporated moisture to provide hot-house conditions.

It was supposed that as the terrestrial heat became exhausted the clouds of vapor broke and climates gradually came under the régime of the sun, with hot regions where the sun was nearly vertical and cooler and cooler ones toward the poles because of the low angle with which its rays struck the surface. Then the world was overtaken by a fearful calamity, the glacial period was ushered in and millions of square miles of Europe and America were covered with ice sheets.

With a waning sun and a chilling earth the final catastrophe was approaching when all life must end for lack of warmth. Within a measurable time the earth would undergo the frigid death of the moon.

The poets were appalled by this cheerless vision of the end of all things that man holds desirable and have painted the despair of the last group of miserable survivors cowering in darkness over the embers of the last fuel before everything went out in everlasting night.

Not much more than a generation ago this gloomy foreboding was generally accepted as justified and the only question to be discussed was the number of thousands or millions of years before the end should come.

It is curious to note that the first lifting of the cloud of gloom came from the discovery of an ancient ice age in India and then in South Africa and Australia. The hothouse conception of the ancient earth sprang largely from the supposed climatic conditions of the Carboniferous forests of Europe and North America, where giant club mosses and tree ferns suggested heat and moisture. The scientific world suffered a shock when it was proved that during, or just after, the period of lush tropical forests extending far to the north in Europe and America, vast glaciers spread over what are now subtropical regions in both hemispheres.

For years there was doubt and hesitation in accepting this revolutionary idea; but at length it became evident to all that the world's most tremendous glaciation had come many millions of years ago toward the end of the Palæozoic. Not long afterwards, as geological periods go, the earth had completely recovered from this Palæozoic refrigeration and had slipped back into its usual mildness reaching to the poles.

The nightmare was over and it became evident that if

the world had not gone out in chill and darkness after the fearful glaciation at the end of the Carboniferous there was no reason to fear for the future. It would probably recover completely from the less intense Pleistocene glaciation from which it now seems to be slowly emerging. The prophets of disaster to come were silenced.

The discovery of the late Palæozoic ice age cast great doubt upon the theory of an earth cooling from a gaseous or molten beginning and strongly supported the uniformitarian view that climates constantly vary but within comparatively moderate limits.

Of late years the evidence for this has become overwhelming. Ice ages or times of cooler climate than the average are known to be scattered all along the geological history of the world, one occurring even in Huronian times; and the earliest rocks include sediments formed under water, showing that the earth was even then not too warm for lakes or seas to exist.

"That which hath been shall be." Conditions permitting of the existence of liquid water have been maintained on the earth through all the hundreds of millions of years since the known geological record began. There have been very long periods of milder climate than the present and a few relatively short ice ages when it was far colder than now; but the water of the earth has never been wholly evaporated during the warm periods nor completely frozen during the ice ages; otherwise the life of the world must have perished, since liquid water is necessary for its existence. In fact the temperature of at least some portions of the earth must have kept within the limits of freezing and about 190° Fahrenheit to allow its inhabitants to survive. That a little globe like the earth, exposed on all sides to the cold of space, should have kept the temperature of its surface within these narrow limits for the half billion or more years since life has existed upon it is a most astounding fact, one that would be quite incredible if the nebular theory of the earth's origin as usually stated was well founded.

Evidences of more or less important times of cooling of the earth have been multiplying within the last few years in a surprising way, and all modern works on historical geology refer to them. Perhaps the most interesting proof of this trend is to be found in Reed's Geology of the British Empire, published in 1921, in which there are suggestions of ancient glacial action on no less than thirty-one pages, without including the many references to the Pleistocene.

Glaciation has been suspected or proved in the Pleistocene, the Eocene, the Cretaceous, the Jurassic, the Triassic, the Permian, the Carboniferous, the Devonian, the Silurian, the Ordovician, the Cambrian, the late Pre-cambrian, the Huronian and the Sudburian or Timiskamian. From this list it is seen that in almost all of the major divisions of geological time ice action has at least been suspected. In most of these times the known marks of refrigeration are on a comparatively small scale, however, and probably imply only local mountain or piedmont glaciers.

At four of these times, the Pleistocene, the Permocarboniferous, the early Cambrian or late Pre-cambrian, and the Huronian, glaciation was on so broad a scale that glaciers of the continental type must have been at work, and hence there must have been a very serious lowering of temperature in the area affected, and probably in the whole world.

When it is remembered that glacial deposits are mostly formed on the land while the geological history of the world is recorded mainly in sediments under water, the showing just mentioned is very impressive. The chances that loose drift materials left by an ice sheet should be buried and preserved under marine deposits before rain and weather and running water have rearranged them past recognition are by no means good.

To this must be added that the record, particularly in the older formations, is incomplete, and that the oldest formations of all are usually so squeezed and metamorphosed into schists that the original structures have been largely destroyed.

It is probable that glaciers of the alpine type have existed on high mountains ever since the geological record began, and one may expect that evidences of such local glaciation will increase as the geology of the world becomes better known.

In spite of the numerous proofs of ice action mentioned above it must be kept in mind that usually the climate of the world has been mild. It has been mentioned that even the Arctic and Antarctic regions once had rank vegetation when trees grew and seams of coal were laid down, and the paroxysms of cold must be looked upon as merely brief interludes between the long periods of warmth. The short ice ages were, however, of enormous significance for living beings, since they were times of fierce testing both for plants and animals, resulting in the destruction of the weaklings and the survival of the vigorous and hardy species. These short spells of trial and stress meant far more for the development of the world's inhabitants than all the long periods of ease and sloth when the earth was a hot-house. After the cold periods the forms that had survived the ordeal rapidly advanced and expanded to occupy the spaces left vacant.

This was true, for instance, of the reptiles which inherited the earth and occupied its waste places after the Permocarboniferous glaciation; and again after the less intense Eocene refrigeration, when the mammals rapidly advanced and multiplied to take the place in their turn of the cold-blooded reptiles which had succumbed to the change of climate.

The climatic changes just suggested seem, when judged by human standards, to proceed with exceeding slowness, so that the thousands of years of the historic period have not sufficed to produce any marked climatic change. It is only as unexpected features in the long perspective of geological ages that glacial periods stand out as sudden and dramatic events in the history of the world. Such events are, however, of peculiar interest to humanity as involving changes in living beings. It may be that the races of civilized men are merely evanescent phenomena bound up with the bracing climates of a brief ice age, to sink, after a few more thousand years, into a state of tropical sloth and barbarism when the world shall have fallen back into its usual relaxing warmth and moisture, the East African conditions which have been so customary in the past.

In the present work the long periods of heat, which make up so much of the world's history, will be left to one side, and the brief episodes of cold, usually passed over lightly, or even unmentioned in books on geology, will be studied in some detail.

Ice ages do not occur with regularity in the geological record. They seem to be spaced in a haphazard way and there is no certain evidence of a rythmic swing, after so many millions of years, from mild to cold conditions and back again. In fact there seems to be something accidental in regard to their arrival and especially as to the regions which shall be ice covered, something incalculable from our present data. We have not yet found a satisfactory explanation why India, Africa, Australia and South America developed great ice sheets at the end of the Palæozoic, while Europe and North America were only lightly touched by frost. In the last ice age these two continents were half covered with ice, while the others almost escaped. There is a complete reversal of conditions when we compare the effects of the two greatest known ice ages, and no good reason can yet be assigned for this reversal.

It has been suggested with a good deal of probability that such refrigerations are associated with diastrophic changes in the earth's crust, times when mountain ranges are elevated and continents emerge from shallow epicontinental seas. Thus barriers may be formed interfering with the circulation of the atmosphere and the ocean, and fresh surfaces will be exposed to weathering. A sufficient elevation of mountains would imply glaciation of their summits long before ice could form on lower ground.

In succeeding chapters a short study will be made of the work of modern glaciers and ice sheets, so that their mode of operation may be understood, and the geological results of their work will be noted for comparison with other glacial deposits. A brief account will be given of the last ice age, that of the Pleistocene, whose deposits can be studied in many of the temperate regions of the world where living glaciers do not occur.

These formerly glaciated regions are really more helpful in the study of ice action on the large scale than the still ice covered areas, such as Greenland and Antarctica, where severe weather conditions and vast mantles of snow make observation of the glacial machinery and of the grist which it turns out very difficult. In the northern parts of Europe and America such material is widely spread and well preserved, so that the effects of ice in scouring rock surfaces and depositing drift materials can be studied much better than in the case of existing ice sheets.

An outline will be given of the distribution of the Pleistocene glaciers; the important question of the occurrence of interglacial periods, times when the ice was removed and then returned again, will be discussed; and the length of time required for these changes will be considered.

Having prepared the way by a study of the comparatively well known Pleistocene glaciation, the more ancient ice ages will be taken up in descending order; since the evidence of ice action is more and more difficult to obtain as one goes downward in the geological scale, and it is desirable to pass from well ascertained phenomena to similar but less legible evidence in more ancient rocks.

Naturally the most tremendous ice age on record, the one near the end of the Palæozoic, will be described at greater length than the others.

In the last chapter the more probable theories of the causes of glacial periods will be discussed.

So far as possible references will be made to the literature of the subject, but of late years descriptions of ancient glaciation have grown beyond all bounds and no one can hope to cover the whole ground.

Living Glaciers and Their Motions

There are many varieties of glaciers, and elaborate classifications have been suggested for them, but for our purpose only three kinds need be mentioned: (a) mountain glaciers, which begin in fields of perpetual snow and flow down some valley until melted by the rise of temperature at lower levels; (b) piedmont glaciers, where mountain glaciers reach the end of their valley and spread out as a bulb or sheet on the more level ground; and (c) continental ice sheets, which are not dependent on mountains and may begin at low levels, spreading widely without much reference to the topography.

Mountain glaciers may be studied in all lofty ranges, even under the equator, and are particularly numerous and accessible in the Alps, the Canadian Rockies and the Alaskan and New Zealand mountains. Piedmont glaciers are best illustrated along the Alaskan coast; and continental ice sheets are found on a large scale only in Greenland and the Antarctic continent, where few geologists have an opportunity to study them. The different types may be associated together, as in Greenland and Antarctica.

The easily accessible mountain glaciers provide examples of the most important results of ice work, the larger continental sheets merely showing the same effects on a vaster scale, covering hundreds of thousands of square miles instead of perhaps a fraction of a square mile.

At first sight the tongue of ice descending from a snow-field into a valley seems a clumsy tool to do geological work. It is a mass of solid and brittle material that appears devoid of all flexibility and even of the power of movement; and yet this seemingly immovable solid has a facility of transport and sculpture that even running water can hardly equal, and it impresses a most unique and striking set of forms upon any region it has occupied. Our most splendid mountain ranges, with their sharp peaks, their deeply hollowed U-shaped valleys and their rock-bound lakes, owe their daring shapes to the work of ice; and in much of northern Europe and North America the lowlands received their smoothly swelling or hilly forms from the shaping of the rocky surfaces and the spreading of the loose materials by ice.

The thousands of years of later influences have not greatly modified the impress the ice sheets left upon the continents.

The engine of ice, though usually moving only a few inches or a few feet in a day, is tireless and singularly well equipped for its work. For the supporting of heavy weights it is a solid and can carry on its back all the masses quarried from the cliffs by frost and rolled down upon its surface by the avalanche. Blocks the size of a cottage are borne away without effort, though very slowly, and for hundreds

Photographed by Melson.

Erratic Block, Alaska Boundary.
xxvii

of feet from its edge the glacier may be buried under the débris heaped upon it.

And yet the whole vast mass of apparently solid ice with the load upon it is in constant slow motion, is constantly changing its form to adjust itself to the rocky surface beneath, and is always advancing in answer to the pull of gravity. The center of the icy stream moves faster than the edges, as in a river of water. The motion is more rapid on the outer side of a curve in its channel than on the inner one, just as in the flow of water; but as it is a brittle solid its mode of adjustment to these changes and to the varying grades of its valley is, of course, entirely different from that of a liquid.

The actual mechanism by which it moves is not entirely certain. Glaciers are made up of "grains" of ice, each an imperfect crystal, which are capable of adjustments among themselves. Pressure lowers the freezing point of water and when, in the motion of the mass, one grain presses too severely on another, a little liquid water is formed which slips away to a place of less pressure and instantly becomes solid again. This property of "regelation" no doubt aids in the slow movements of the glacier, but the process goes on so inconspicuously as not to be observed. On the other hand there are large scale adjustments of the ice, forming "crevasses" and even "seracs," which are among the most striking and important of their features.

Where the slope of its bed suddenly grows steeper the advancing glacier simply breaks across in fissure after fissure, gaping at the top and narrowing to nothing at the bottom, forming crevasses. As the center of the ice moves on faster than the sides, which lag because of the friction of the rocky edge of the channel, there are less regular crevasses formed, reaching diagonally toward the almost motionless edge. The whole surface of the ice becomes split

Photographed by P. W. Green.

Ice Fall and Seracs, South Dawes Glacier, Alaska Boundary.

up under these conditions by crossing crevasses, and sharp-edged blocks and pinnacles may result, the seracs that make so much trouble to the mountaineer.

From a distance this splitting up of the ice looks like the foam of a mountain torrent and suggests violent motion rather than the deliberate advance of a foot or less in a day.

Once past the obstruction or change of grade the ice masses weld together again and present an unbroken surface as uniform and solid as before; and the whole effect of these adjustments among glacier grains, crevasses and seracs is that the glacier moves downwards like a plastic substance, such as pitch.

Beside the visible surface adjustments of the moving ice which have just been described there are shearing motions near the bottom of the glacier, due to the friction of the floor, the lowest layer moving very slowly and the higher ones more and more rapidly, like a viscous material, and so things enclosed in the. lower levels may be carried up to higher parts of the glacier.

How Glaciers Scour Their Valleys

Having an idea of the motions of the ice one may next consider the kind of geological work accomplished in advancing over the rocky floor beneath. That boulders and finer materials are easily carried on its surface is evident; and also that if crevasses open beneath the load, rock fragments will slip into them and may be caught in the narrowing of the fissure or may actually reach the bottom. When the fissure closes in the onward motion such blocks are firmly held as tools to grind and striate and groove the rock surface beneath, while finer particles serve as polishing powder to give a smooth finish to the work. Any harder projecting ribs of rock will be rounded on the side from which the ice comes and form *roches moutonnées,* and sheltered behind

Débris Laden Glacier. Mt. Robson, Canadian Rockies.

them there may be lodged part of the load of rock fragments dragged along by the moving ice. The direction of
the striæ and the position of the masses left in the lee of the
projections show the line of motion of the ice. Sometimes
this direction changes and then there will be later striæ

Roche Moutonnée, Thousand Islands, Ont.

partly obscuring the earlier ones, hinting at a complicated
history of the glaciation.

Where the floor beneath the glacier is covered with loose
blocks, resulting from weathering in an earlier period of
warmer climate, tools are already at hand for abrading projecting points; and even unweathered rock, when broken
up by joints, may have fragments plucked out and carried
along to carve down the bed of the glacier.

The smoothed and striated surface left by the action of

ice is one of the most satisfactory proofs of glaciation, but near the edge of a glacier, where its lower part is overloaded with débris gathered in the ways described, its motions become more and more sluggish, until finally the bottom layers completely stagnate and the less encumbered overlying ice passes over them without touching the rock floor. In such cases there will be no striated surface, and in the work

Striated Stone, Main Glacier, Mt. Robson.

of great ice sheets even thousands of square miles may thus escape the scouring which is so striking a feature beneath the more active part of the ice.

Striated Stones and Boulder Clay

The tools which have served to polish and striate the rocks over which the glacier moves are themselves abraded, rounded at the edges, facetted, polished and striated, and their faces are frequently scored in more than one direction.

Recent Boulder Clay, Cora Island, Spitzbergen.

xxxiv

Medial Moraine, Alaskan Boundary.

Such "soled" and striated stones are very characteristic results of ice work. They are typical manufactured articles: shaped and marked in unmistakable ways as the handiwork of craftsman ice. Their blunted edges and smoothly carved forms, sometimes with inward as well as outward curves, are directly related to the physical properties of ice, which held them firmly under strong pressure and yet permitted readjustments and differential motions, as layer was thrust over layer, so that fragments might change their setting and be ground on different sides. The results are unmistakable, since no other process of nature gives similar shapes or similar surfaces. Typically glaciated stones are the best evidence of ice action, since they cannot be closely imitated by any other agency.

While the stones undergo the shaping process they are, of course, being worn down, and the rock beneath is also ground away, so that a large amount of "rock flour" results; and, in fact, particles of all sizes from the finest dust to boulders several feet in diameter and tons in weight come as grist from the glacial mill, and when the ice finally melts the whole mixture of unassorted materials is left as "till." Most rocks when ground up make an impure clay, so that till is often called "boulder clay," though there are also sandy varieties where siliceous rocks have furnished the materials.

An ancient till which has been consolidated to rock is called "tillite"; and tillites with their soled boulders and striated stones are the best proofs of ancient glaciation.

A striated surface beneath confirms the evidence, but its absence is no argument against the glacial origin of a boulder conglomerate having the character of a tillite, for reasons which have already been mentioned.

It should be added that in the central area of an ice sheet, as well as near its edge, there may be few evidences of

Photographed by E. S. Moore.

Terminal Moraine and Kettle, Variegated Glacier, Alaska.

xxxvii

ice action, since there will be little or no scouring of the surface beneath, and even the ancient weathered materials may remain almost undisturbed though powerful erosion is going on all round the unmoved center from which the ice radiates.

The features just described are usually the most important for the student of glacial periods, but certain marginal phenomena may be of interest, also. There may be terminal, lateral, medial and interlobate moraines, heaps or ridges of boulders with comparatively little clayey or sandy matrix. As a rule few polished or striated stones are found in moraines, since the materials have been transported on the surface or in the ice itself and have not served as grinding tools at the bottom.

There may also be glacio-fluvial deposits, such as kames and eskers, partly stratified by the action of water, though these have seldom been described from ice ages older than the Pleistocene. In these, also, striated stones are rare or absent.

Seasonally Banded Clays

In many cases glacial rivers have deposited seasonally banded clays in adjoining lakes, the *varve* clays of DeGeer and other Scandinavian geologists. These are sometimes called "ribbon" clays. They are composed of annual layers having a coarser portion laid down during summer and a finer portion representing the slow settling of the smallest particles in winter. R. W. Sayles has described such banded clays associated with tillite in various places in America, and David and Süssmilch have given an account of them in connection with glacial beds in New South Wales. They are corroborative evidence but not independent proof of glaciation, since similar seasonally banded deposits may be formed in temperate climates having cold winters.

Varves in Pleistocene Clay, Toronto, Ont.

Metamorphosed Glacial Deposits

Where undisturbed even very ancient glacial deposits show the characteristic features that have just been mentioned, and unmistakable tillite with well striated stones occurs as far down as the Huronian of Ontario; but in many regions of mountain building the rocks have undergone so much squeezing, crushing and shearing that striated surfaces have been destroyed and often the whole rock has become schistose so that there is not even the appearance of tillite.

Even under those conditions there may be features that prove, or at least strongly suggest, glacial action. If a boulder conglomerate contains angular, subangular and rounded blocks of large size and of kinds that must have come from a distance there is a strong probability that the deposit is glacial; and if there were no mountains near by at the time of formation to permit of landslips or mud flows or the sweep of torrents, the deposit is almost certainly the work of ice.

An unweathered and gritty matrix also is corroborative evidence of a cold climate. The matrix of tillites is usually of the nature of graywacke or arkose and generally contains fresh and angular fragments of feldspar.

Glacial deposits often vary greatly in thickness, structure and composition and this variability itself is suggestive of ice work. The till may consist of unstratified stony clay passing into banded, almost stoneless, varve clays; or there may be beds of stratified clay or sand or gravel between two layers of till; or a bed of till may pass into the coarse stratified materials of a kame or into the still more bouldery, but unstratified mass of a moraine. All of these features are suggestive of glacial action and even without the finding of striated stones may combine to make it certain.

Dwyka Tillite, Prieska, South Africa.

The extent of such bouldery deposits is of great importance, since a small morainic mass or a limited area of tillite may mean only a local mountain glacier, in places where the topography of the time permitted. A wide area of tillite, especially if beds associated with it carry marine fossils, can be interpreted only as the result of large scale

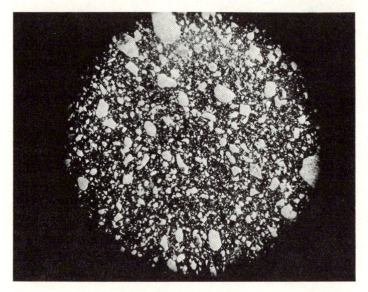

Matrix of Dwyka Tillite.

refrigeration, i.e., as the work of a definite lowering of temperature; and where tillites are found at the same horizon on several continents the refrigeration must have reached the intensity of an ice age and must have implied a catastrophe changing the whole life conditions of the plants and animals of the time.

It will be understood, of course, that the evidence for an ice age will be more and more difficult to find in the older formations, not alone because of metamorphic changes obscuring the record, but also because large areas of the tillite must have been completely destroyed or must have been buried under later formations so as to be out of reach of geologists.

In the most ancient periods of the earth's history lack of evidence of glaciation does not necessarily mean that no glaciation took place, because of the fragmentary preservation of the record.

PART I

THE PLEISTOCENE ICE AGE

CHAPTER I

General Features

THE Pleistocene Ice Age is naturally the one of which we know most, since the earth is only now recovering from its effects, if indeed we are not living in an interglacial period; and the drift deposits which tell its strange story cover millions of square miles in countries occupied by some of the most highly civilized nations.

The investigation of glacial matters is one of the more recent sides of geology; yet a vast literature has already sprung up in regard to the Glacial Period, as it is commonly called, and all intelligent readers are more or less familiar with the subject. Excellent and very readable books on the Glacial Period have been prepared by James Geikie in Scotland, "The Great Ice Age"; F. G. Wright in America, "The Ice Age in North America": and W. B. Wright in Ireland, "The Quaternary Ice Age"; and innumerable articles on this startling event in the world's recent history have been published in scientific journals and also in more popular periodicals. Ancient ice ages have hitherto received little attention in books on the subject. In the present work it is not intended to take up the Pleistocene ice age in great detail, but rather to outline its extent, to describe its mode of operation and to study particularly such features as will throw light on more ancient, and therefore less completely recorded, glaciations. The question of the duration of the Pleistocene glacial period will be discussed, and there will be a

3

consideration of the climatic changes involved in such a catastrophe, of the magnitude of the ice sheets and the importance of the physical effects upon the regions covered. The results of such widespread glaciation upon the life of the world will be referred to also.

The criteria for the recognition of ice action suggested in the previous chapter will apply, of course, to the Pleistocene as well as to more ancient glaciations.

The Drift

In North America the term "drift" covers all the deposits formed by the Pleistocene ice sheets, even when modified by the work of running or standing water; it includes boulder clay, moraines, kames, eskers, outwash aprons, stratified materials in glacial lakes, and also interglacial beds formed under milder conditions. In England the word is used in the plural, "drifts," and the usage is justified by the variable character of the deposits; but as this account is written in America the singular form of the word will be employed.

Mapping the drift in its protean forms enables us to outline the extent of the ice-covered regions, and a study of the materials left by the ice gives valuable information as to the directions from which the ice came. There are, however, often large areas which have been swept bare of loose materials, but even then the striæ and the *moutonnées* forms of the bed rock are eloquent evidences of the work done and the line of march of the invading glacier. It is readily understood that the drift is very unequally distributed, varying from nothing at all, where the scouring was powerful, to thicknesses of 600 or 700 feet where the overloaded ice at the margin of the sheet was dumping its burden for hundreds or perhaps thousands of years.

Before the Ice Age there must have been a great thickness of weathered rock overlying the unchanged formations

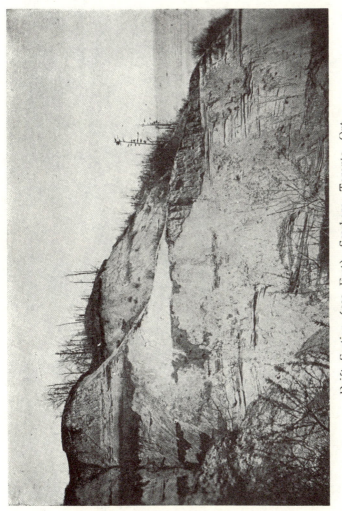

Drift Section (200 Feet), Scarboro, Toronto, Ont.

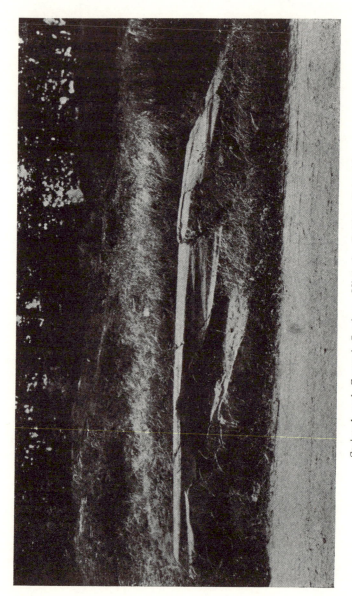

Striated and Gouged Surface, Victoria, British Columbia.

beneath, such as we now see in unglaciated regions.   In most places this has been swept off by the ice sheets, and often, as just mentioned, the underlying surface has been pared down into the solid rock.   The contrast between the bare glaciated hills of Northern Canada and the deep residual soils of the southern United States illustrates this excellently; and between the two types of landscape come the heaped up morainic hills of the margin of the ice sheet.

In the center of an ice sheet little work is done.   Surrounding it is a broad belt of scouring and erosion where load is being gathered, and near the margin the melting ice is depositing its load, but there is no sharp boundary between the different types of surface left by the vanishing glacier.

The drift covered areas are found chiefly in the northern hemisphere because the northern continents provided great land surfaces in sufficiently high latitudes for ice sheets to form.   In the temperate parts of the southern hemisphere land occupies much less space than in the northern, so that the parts exposed to Pleistocene glaciation are relatively small.

Outside of the Antarctic continent, which is still covered with ice, only Patagonia shows a drift-covered area of some importance.

Extent of Pleistocene Glaciation

In earlier times it was supposed that during the Glacial Period a vast ice cap radiated from the north pole, extending varying distances southward over seas and continents. It was presently found, however, that some northern countries were never covered with ice, and that in reality there were several more or less distinct ice sheets starting from local centers and expanding in all directions, north as well

Polished and Striated Surface of Limestone, Dundas, Ontario.

as east and west and south. It was found, too, that these ice sheets were distributed in what appeared to be a capricious manner. Siberia, now including some of the coldest parts of the world, was not covered, and the same was true of most of Alaska and the Yukon Territory in Canada; while northern Europe, with its relatively mild climate, was buried under ice as far south as London and Berlin; and most of Canada and the central and eastern United States was covered, the ice reaching as far south as Cincinnati in the Mississippi Valley.

About 2,000,000 square miles of Europe and 4,000,000 square miles of North America were glaciated, and in addition many mountain areas in various parts of the world, as well as most of Patagonia in South America.

The two great ice-covered regions still remaining, Greenland and Antarctica, make up nearly another 6,000,000 square miles, so that, all told, there were more than 12,000,-000 square miles of the earth's surface glaciated during the Pleistocene, one-fifth of the total land surface of the globe. It is probable, however, that the whole of this area was not covered at one time.

When one remembers that during most of its history the earth, as shown by fossils, has been free from ice sheets even in arctic regions, while now there are about 6,000,000 square miles ice covered, it is evident that we have only half emerged from the Pleistocene glaciation. Until Greenland and Antarctica are set free we cannot claim that the ice age is entirely over.

North America was, after Antarctica, the continent most widely glaciated. There were three great ice sheets, a Cordilleran, a Keewatin, and a Labrador sheet, which at the time of maximum refrigeration more or less coalesced. In Europe there was but one great ice cap, which may be called the Baltic or Scandinavian sheet.

Differences in Climate During the Ice Age

It is interesting to note that there were climatic differences of the same order as at present during the world-wide refrigeration. The depression of temperature seems to have been of the same amount all over the world so that the relative standing of the different parts remained unchanged.

At present the climate of eastern North America is much cooler than that of corresponding parts of Europe. Arctic conditions prevail on the Labrador coast down to latitudes which in Europe include Christiania, Glasgow, Edinburg and London, where the winters are comparatively mild. The Labrador coast is treeless and has eight or nine months of winter just across the Atlantic from the south of Scotland where the laurel and the rhododendron and the ivy thrive in the open. The harbor of Quebec is closed by ice for five months in the year, while that of Glasgow, more than nine degrees farther north, is always open. In the ice age New York, in lat. 40° 30′, was covered, while London, in lat. 51°, was south of the European ice front.

A very similar relation is seen in comparing eastern and western America at present and in Pleistocene times, the mild Pacific climate limiting the southern boundary of the Cordilleran sheet to about lat. 48°, while ice in the Mississippi Valley reached 37° 30′. The differences between the two sides of the Atlantic and the two sides of North America were much the same as at present, showing that the circulation of air and water has not greatly changed since the Pleistocene.

The Cordilleran Ice Sheet

Of the three great continental glaciers in America, that of the Cordillera, or western mountains, probably came first and may be described as typical for glaciation in a mountainous region.

We may assume that during the slow cooling of Pliocene times, local glaciers formed on the higher mountains of the

After W. C. Alden.

Map of Glaciation in North America.

Coast Range, the Gold Ranges, and the Rocky Mountains proper. As the more intense cold of the Pleistocene came on these local glaciers grew longer and more massive and the snowline was progressively lowered until at length the bottom of the main valleys was reached, where the ice spread out forming piedmont glaciers as now in Alaska.

The piedmont glaciers coalesced into broader sheets and were thickened by the inflow of ice from the surrounding mountain ranges until the valleys were filled to the brim and the interior tableland of British Columbia was covered. Thus a great mass of ice accumulated over the whole of British Columbia and the southern part of Yukon Territory, hemmed in on the northeast side by the main range of the Rockies and to a less extent on the southwest by the Coast Range.

In the central parts the ice reached an elevation of about 8,000 feet but, owing to the arrangement of the mountain ranges, could escape freely only toward the northwest, where it feathered out in the southern part of Yukon Territory, and toward the southeast, where the great mountain trough passed into the United States. The high wall of the Rockies prevented it from reaching the plains except on a small scale through Athabasca Pass and Bow Pass. The lower barrier of the Coast Range gave outlets of some importance through the valleys of the Stikeen, Skeena and Fraser rivers. How far the ice encroached on the Pacific is not known, but the channel between the mainland and Vancouver Island was filled and the Queen Charlotte Islands were covered, so that there was a great extension beyond the Coast Range on the southwest.

Although the Cordillera as a whole was flooded with ice, all the higher mountains rose above the waste of snow and glacier as "nunataks," to use the Greenland term; since transported blocks are not found above 8,000 feet. The loftier mountains must have appeared like islands rising out of a sea of white.

The area covered with ice was about 1,200 miles long, running parallel to the Pacific coast, and from 250 to 400 miles broad, and included not less than 350,000 square miles, while isolated glaciers on a large scale occurred farther to the northwest, along the Alaskan mountains, and to the south in the Olympic and other ranges of Washington, Oregon, Idaho and Montana.

The ice was at some points a mile thick and so vast a mass might be expected to do an immense amount of work; but in reality it accomplished comparatively little. The lower, narrowly enclosed part of the ice, once the valleys were filled, became stagnant and left scarcely any impression on the country, even the placer gold deposits of preglacial origin being preserved in many places.

Wherever it moved out through narrow valleys or passes, however, there was powerful glaciation and the rocks have been polished, striated and shaped into *moutonnées* forms. The main outlet valleys have been carved into U shapes easily recognized even from a railway train.

The Cordilleran ice sheet did its chief work around its margins. In the interior valleys and the central tableland of British Columbia often very little evidence of ice action is found, and one must climb the mountains to find boulder clay, erratics and striated surfaces. The currents of ice were to a large degree superficial, the lower level remaining almost quiescent.

Except that its marginal mountains are in general lower, Greenland at the present time presents a close analogy with Cordilleran conditions in the Pleistocene. The whole interior is covered with perpetual ice through which rise nunataks near the edge, where the mountainous rim is broken in places by valleys or fiords through which effluent glaciers escape to the sea.

The Keewatin Ice Sheet

Probably after the Cordilleran ice had reached its full size, but long before the mountain glaciers separated and shrank to their present dimensions, a vast *neve* accumulated to the west of Hudson Bay on comparatively low ground, perhaps averaging 1,000 or 1,500 feet above the sea, with no mountains, and therefore no Alpine glaciers, as a starting point. The Keewatin *neve* expanded in all directions and finally reached the foothills of the Rockies toward the west, the Missouri and Mississippi toward the south, the Lake-of-the-Woods on the east and some of the Arctic islands on the north. It encroached on Hudson Bay, also, but its northern and eastern boundaries have not been definitely worked out.

The appropriate name of the ice sheet was given by J. B. Tyrrell, who first showed that there was a separate ice sheet west of Hudson Bay, and who has done admirable work in outlining and describing this tremendous glacial mass, which before his time was included as part of a supposed Laurentide ice sheet covering the Laurentian region of Canada.

The Keewatin ice sheet extended 2,000 miles from north to south and 1,300 miles from east to west, and its area when most widely extended must have exceeded 1,500,000 square miles. It seems to have feathered out in all directions, unhampered by any important barrier, except perhaps the foothills of the Rocky Mountains, which it just touched. It was a typical example of a continental ice sheet, with no complications except its relations to the Labradorean sheet, its neighbor toward the east.

The most extraordinary feature of this ice sheet is the fact that it transported Archæan boulders from the Laurentian area near Hudson Bay to the foothills of Southern

Alberta, depositing them at a level of 4,500 feet, as shown by G. M. Dawson. Crossing the prairies one sees red granite and gneiss boulders and others of green schist scattered over the surface as far as Calgary and beyond Edmonton. Some of these boulders are more than 500 miles from the nearest known source and have been elevated at least 3,000 feet above their starting point.

Dawson accounted for this by supposing that the land toward the southwest was much lower than now, even at sea level, but there is no evidence in the way of fossils to prove this; and the relative levels of the edge and middle of the Keewatin sheet were probably not very different from those of the present. It is, in fact, very likely that the glacial center, being more heavily weighted with ice than the edge, was some hundreds of feet lower than now during the time of greatest glaciation, so that the ascent was really more than 3,000 feet.

This illustrates a marked difference between the transporting power of ice and that of water—it can carry its load uphill, whereas water can only carry it down.

After the statement made earlier that mountain glaciers flow down their valleys by the pull of gravity, it may naturally be asked how an ice sheet can reverse the process and move up grade. This is accounted for by the fact that the inclination of the surface of the ice and not the slope of the land beneath determines the flow. If the surface of the Keewatin sheet sloped for 500 miles toward the southwest, the transport of blocks of granite to that distance with an elevation of 3,000 feet on the way can be accounted for.

There were no Archæan nunataks rising above the Keewatin ice sheet, so that the boulders found in the west must have been carried englacially, embedded in the lower ice layers, and could not have been borne on the surface like morainic materials on mountain glaciers. We must imagine the ice sheet clogged with débris near its edge, but covered with pure white snow in the interior like the surface of the inland ice of Greenland.

The Labradorean Ice Sheet

The Labrador ice sheet probably began later than the Keewatin sheet but was in existence when the latter had reached its maximum dimensions and then coalesced with it. The Labradorean glacier ultimately covered almost the whole of eastern Canada and extended south to Cincinnati and New York in the United States. Its center was in northern Quebec to the east of James Bay, and the name Quebec sheet would have been more appropriate.

The Labrador and Keewatin ice sheets combined to cover an area of at least 3,500,000 square miles, and for a considerable time the basin of Hudson Bay was occupied by ice. To the west of Lake Superior the two sheets seem to have jostled one another, sometimes one and sometimes the other gaining the upper hand, as shown by the distribution of boulders from one or the other quarter, and also by the occurrence of two sets of striæ, one coming from the north and the other from the northeast, the latter being the more recent. The extent of the overlap of the two drift sheets has not been worked out very completely. A similar struggle has been shown to have taken place in the Hudson Bay region.

The thickness of the central parts of the Labrador sheet is unknown, but in northeastern Labrador it was not more than 2,000 feet, since glaciation is not found above that level on the mountains. A large area of the Torngat tableland was left unglaciated, though important local glaciers carved valleys down to sea level on the northeast.[1] In Newfoundland it was not much more than 1,000 feet thick and thousands of square miles of tableland were untouched by it, while in southern Quebec the ice reached about 2,500 feet and left uncovered the higher parts of the Shickshock mountains.[2]   To the south it nearly reached the highest points of the Adirondack mountains, about 5,000 feet above. the sea.

Two important facts have been established by Low, who worked over the central parts of the Labrador sheet: first, the center of glaciation shifted its position, at one time being in lat. 51° or 52°, later in lat. 54°, and finally in lat. 55° or 56°. Instead of beginning at the north and growing southward it reversed this direction; second, that the central area shows few signs of glaciation, so that the preglacial débris due to ages of weathering are almost undisturbed. A broad circle around it is scoured clean to the solid rock, which is *moutonnées* and polished and striated in the most typical way, as can be seen in the bare rocky hills of northern Ontario.

One must think of the great ice sheet as having a broad central cone of stagnant ice, like that in the interior of the Cordillera, while all around this it was effectively denuding the surface and transporting the materials outwards, where thick sheets of boulder clay as well as moraines were being deposited. Near the outer edge of the sheet typically scoured surfaces are to be found only exceptionally, where the ice was forced to ascend the rocky border of a basin, as south of the Great Lakes, or where it crowded through some ravine or narrow valley between rocky hills.

In concluding this account of the three great ice sheets it should be mentioned that there were also minor sheets, like the Patrician sheet in northwestern Ontario, described by Tyrrell,[3] which became merged into the vast combined sheet of the Keewatin and Labrador ice; and that there were many local glaciers on the mountains near the border of the Labrador sheet. There were also much smaller ice sheets on some of the great Arctic islands of northern Canada. One of them still persists in a stagnant state on the tableland of Baffin land, and there are also thousands of square miles of ice cap on Ellesmere and Grinnell lands still farther north.

It is believed that the three great glacial centers were occupied successively, the western one first, then the central one, and finally the eastern, which certainly was the last to disappear. It will also be observed that the Labrador ice sheet was the most southern of the three. In comparing the Keewatin and Labrador sheets their relative positions correspond to the present arrangement of the isotherms, though greatly shifted toward the southward.

There are puzzling features in regard to the successive appearance of ice sheets from west to east. It is natural, of course, that the Cordilleran sheet should come first, since the mountains, by their elevation, were the seat of local glaciers which could expand as the cold increased; but why should the Keewatin glacier begin before the Labrador glacier? Again, why should the fourth great ice sheet, that of Greenland, persist after the Labrador sheet ended? Did the Greenland sheet commence last of all, and is it now on the wane, and will it disappear entirely in some near geological future?

There are, of course, two factors entering into the formation of an ice sheet—a low enough average temperature and a sufficient supply of moisture to fall as snow. Cold alone will not cause glaciation, as is proved in the case of Siberia which was unglaciated, though in parts it is the coldest region in the world outside of the Antarctic continent.

One might imagine that the presence of an ice sheet so influenced the storm tracks, and therefore the deposit of snow, that successive areas were invaded on the southeastern side of the already glaciated area.

Conditions South of the Main Ice Sheets

To the south of the glacial boundary there were mountain glaciers wherever peaks reached a sufficient height. They existed on some of the Appalachian summits and occurred in great numbers and often of large size on the Rockies of the western states. None have been reported from the Mexican mountains, however.

Accompanying the formation of mountain glaciers there was a great change in the climate of the western region, which is now arid or semi-arid, but then had plentiful moisture. Valleys now dry, or containing only salt lakes without an outlet, were then well watered and contained fresh-water lakes.

The best example of the kind is the great lake named by Gilbert Bonneville, which filled to the brim the arid basin of Utah and overflowed to the northward. Great Salt Lake is a shallow remnant in which salts have been concentrated, but hundreds of feet up on the sides of the basin the old shores are still perfectly preserved.[4]

Lake Lahontan had a similar history,[5] and one may speak of the Glacial period of the north as having a parallel in the Pluvial period in the dry states to the south of the ice boundary.

Interglacial Periods in America

The Pleistocene ice sheets were, no doubt, constantly changing their boundaries to correspond to varying temperature and precipitation, just as the modern glaciers of the Alps have been doing since historic times, and the question naturally arises as to the extent of such variations. Were there merely retreats and advances of a moderate kind, or were the ice sheets ever completely removed and, after an interval, restored again? Were there real interglacial periods with a mild climate between glacial periods in the Pleistocene?

No question in regard to the history of the Pleistocene has been debated more earnestly than that of interglacial periods, and geologists are not unanimous as to the answer that should be given to the question. That there have been times of retreat when vegetation and animal life covered marginal parts of the region; and that the glacier has buried their remains under boulder clay during a later advance are admitted by all students of the drift of North America; but there are a few geologists who consider that a comparatively short withdrawal of the ice would account for all the facts. They think of the ice age as a unit unbroken by any important change of climate. On the other hand, a majority of the students of the North American Pleistocene believe that the ice was completely removed during one or more interglacial periods, so that the Pleistocene included an alternation of glacial and temperate climates.

It might be supposed that so important a change would leave behind it evidence that no one could dispute, and that there should be no room for doubt as to what happened in so recent a time of the earth's history. In reality the proof of the complete disappearance of the ice and its return at a later time is, in the nature of things, a matter of great difficulty and it is not surprising that there are differences of opinion on this important question.

The land deposits formed in an interglacial period will consist of soils and peaty materials, while water deposits may include marine beds or sheets of stratified clay, sand or gravel laid down by lakes or rivers. Ordinarily these materials will be thin and quite unconsolidated. The ploughshare of the readvancing glacier must pass over them and in the great majority of cases they will be swept away, ground up and mingled with the next sheet of boulder clay.

Fragments of wood or bits of shell embedded in the till may be all the evidence that remains, and they may be explained as preglacial instead of interglacial.

The only parts of a glaciated region where one may hope to find interglacial deposits widely preserved are marginal, where the overloaded ice is dumping its burden. There even soft beds of peat or sand may survive and be preserved. Such deposits are found in the row of states along the ice border, and there are many reports of logs of wood or old soils or beds of peat between sheets of boulder clay within a hundred miles of the margin. Farther to the north, however, such beds will be preserved only here and there in some river valley or lake bed which the ice has not scoured out to the bottom. Such conditions are rare, and in the wilderness of northern Canada are likely to be overlooked.

The absence of known evidence of interglacial periods near the glacial centers in the Keewatin and Labrador areas is, then, quite to be expected, and is not to be looked upon as conclusive evidence against them.

In Iowa and other states in the Mississippi Valley an elaborate classification of drift sheets has been worked out, as follows:

> Wisconsin Till Sheet
> > Peorian, Interglacial
>
> Iowan Till Sheet
> > Yarmouth Interglacial
>
> Illinoian Till Sheet
> > Sangamon Interglacial
>
> Kansan Till Sheet
> > Aftonian Interglacial
>
> Nebraskan Till Sheet.

The later till sheets overlap the older ones and there is a very marked difference in the amount of weathering of

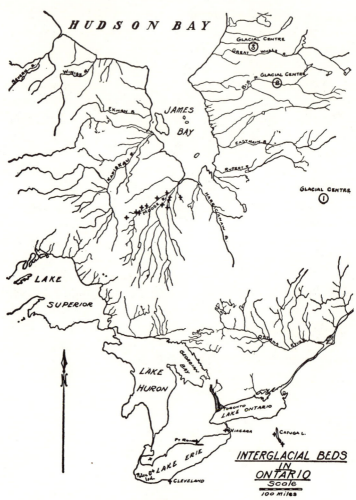

Showing Interglacial Beds in Ontario.

the different sheets, indicating an important lapse of time between them. Interglacial soils or peat, and beds of "gumbotil" or greatly leached boulder clay, as well as loess, are found in the intervals.[6, 7] Except that there is some doubt as to the importance of the Iowan ice sheet, most glacial geologists in the United States accept the classification as valid for the Mississippi Valley, and believe that there was a widespread retreat of the ice in each interglacial time. Since, however, the glacial centers were hundreds of miles north of the boundary the evidence from the United States is not conclusive as to the complete removal of the ice in the interglacial periods.

The Toronto and Moose River Interglacial Formations

In Canada, where the final solution of the problem must be looked for, the number of geologists who have made detailed studies of glacial matters is not large, yet interglacial deposits have been found in British Columbia, Alberta, Manitoba and Ontario; and in the last named province the evidence for at least one complete removal of the ice from the Labradorean center is unmistakable.

The most interesting and extensive deposits of interglacial age have been described at Toronto [8] and along tributaries of Moose River [9] in the James Bay region.

The Toronto Interglacial Formation consists of delta materials brought into a great lake by an important river flowing from the north, probably draining the valley of the present Georgian Bay and Lake Huron. The delta deposits rest on three or four feet of boulder clay formed by an early Labradorean ice sheet and including boulders of Archæan rocks whose nearest outcrop is 90 or 100 miles to the north. At the Don Valley brickyard about twenty-five feet of highly fossiliferous stratified clay and sand rise above the boulder clay.

The species collected include 41 shellfish, 42 plants, mostly preserved as leaves but some as logs of wood, and 7 mammals. The plants are mostly trees, such as red cedar, hickory, pawpaw, osage orange, oaks, elms and maples, two

Interglacial Beds 25 Feet Thick. More Than a Hundred Species of Pleistocene Fossils Have Been Found in Them, Including Leaves and Trunks of Warm-climate Trees, Many Fresh-water Shells and a Few Mammals.

of the latter extinct; and several of them belong to more southern localities than Toronto at present. Palæo-botanists and foresters state that the flora corresponds to a latitude four or five degrees south of Toronto, such as Pennsylvania or Ohio. The shellfish also include some species not now found in the Ontario waters but still living in the Mississippi.

Above the highly fossiliferous beds there are stratified clays containing wings of beetles, 70 of the 72 species known being extinct. At the Don Valley brickyard twenty feet of the clay contained 672 annual layers. The thickest section of the Toronto Formation is found at Scarboro' Heights, ten miles east of the Don, where 185 feet have been measured between the two boulder clays.

After the delta was built the great lake, which rose 150 feet higher than Ontario, was drained, and valleys were cut through the delta beds, one of them to the depth of 160 feet. At another place a valley was sunk 16 feet into the solid shale beneath the Pleistocene.

These valleys have gently sloping sides and are much more mature in appearance than the present Don valley, indicating that the latter part of the interglacial period (after the great lake was lowered) was much longer than the whole of postglacial time.

Finally the ice returned and the interglacial delta with its valleys was buried under a second sheet of till, which was followed by three other retreats and advances of the ice, as proved by beds of stratified sand and clay between sheets of boulder clay. The formation has been shown by borings for town water supplies to extend as an old channel to Barrie near the Georgian Bay, a distance of 64 miles.

Beyond Barrie toward the north for 350 miles interglacial deposits have not yet been discovered; but on Moose River and its tributaries, not far from the southern end of James Bay, such beds have been found at many places, including sections seventy feet thick of stratified clay with peat and wood like that of the Don and Scarboro. Trees eighteen inches in diameter grew there at the time, so that the climate must have been at least as warm as at present. Marine shells have been found in one of these interglacial deposits three hundred feet above the sea, indicating great changes of level not unlike those which caused the damming of the interglacial lake at Toronto.

The interglacial beds of Toronto have not been traced through to the Mississippi Valley, though similar fossils have been found south of Lake Ontario, at Niagara, and on the north and south shores of Lake Erie; so that it is uncertain which of the interglacial beds to the southwest corresponds to the Toronto formation. The large number of extinct beetles and mammals, and the fact that four sheets of till with unfossiliferous beds between them overlie the Toronto deposits indicate that the Toronto formation belongs far back in Pleistocene time, so that it is perhaps of the same age as the Sangamon interglacial deposits of the southwest, or possibly equivalent to the Aftonian, from which similar trees and mammals are reported.

The ice retreated after the Nebraskan or Kansan advance for 800 miles toward the northeast, to a point not more than 300 or 400 miles from the Labrador glacial center, east of James Bay, nor more than 500 or 600 miles from the radiant center of the Keewatin sheet toward the northwest. The ice front had passed beyond the height of land between the Great Lakes and Hudson Bay, and both glacial centers were on low ground with no tableland or range of mountains to provide snowfall.

The thickness of the Toronto interglacial beds, and the deep interglacial valleys carved in them by rivers indicate a length of time at least three times that which has elapsed since the last ice sheet set free Niagara Falls. There is no ice now at either the Keewatin or the Labrador center; what is the probability that ice remained east of James Bay for 75,000 years with a warmer climate than at present?

The facts in regard to the interglacial time recorded in Ontario, as given above, seem to prove conclusively that in at least one interglacial period North America east of the Rocky Mountains was entirely freed from ice.

In the territory covered by the Keewatin ice sheet interglacial beds has been reported by Tyrrell as being 70 feet thick and containing plants and fresh-water shells at Rolling River in Manitoba, and similar beds with lignite are known from Rosebud Creek and Belly River in Alberta, nearer the edge of the drift.[10]

In the Cordilleran region interglacial beds have been found near Vancouver, containing plants indicating a mild climate; [11]   and still warmer conditions are proved by leaves of trees and other plants found on St. Mary's River in southeastern British Columbia.

On the plateau of the state of Washington, to the south, three ice advances and two interglacial periods have been reported by J. Harlan Bretz,[12] and an interglacial time has been proved in studies of the mountain glaciers of Montana [13] and Colorado, where there are two sets of moraines, an older, much weathered one, and a later one which is better preserved.

The pluvial period in Utah, when Lake Bonneville filled the valley of Great Salt Lake, was divided into two parts by an arid time when the water was evaporated, leaving deposits of various salts. This arid interval corresponds to a long interglacial period.

From the instances just given of interglacial deposits or of interglacial weathering occurring from east to west across North America there can be no doubt that in at least one mild interval the continental ice sheets completely disappeared and the Cordilleran ice cap was reduced to separate glaciers, while the glaciers of the mountains to the south were greatly diminished, and a warm and arid climate interrupted pluvial conditions in the interior basin of the Rockies. Another widespread interglacial period, though not so mild and of shorter duration, seems very probable; and there were two additional retreats of the ice of some importance in the Mississippi Valley, as shown by the successive drift sheets of Iowa and adjacent states.

The Pleistocene ice age was far from being a time of continuous unbroken refrigeration in North America.

Glaciation in Greenland

As mentioned on a former page, Greenland is still in the ice age, more than three-fourths of its surface being covered with a great central sheet, leaving only coastal fringes to grow green in the short summer so as to justify its name. The uncovered parts are on the southeast, the west and the north of the great island. There is evidence in the form of smoothed rock surfaces and boulder clay and moraines to show that the ice once covered much more territory, though parts at least of the northern area were never glaciated, as I am informed by Dr. Lauge Koch.

It has been shown by American, Norwegian, Danish and Swiss explorers that the ice rises rapidly at the edge, where mountains often lift themselves island-like above the surface as nunataks. Toward the interior the slope is much more gentle, forming two very flat domes, the northern and higher one reaching 3,000 meters in elevation.[14, 15] There appear to be two centers of glaciation which have become confluent.

The greater part of the ice margin is on the land, where thawing balances the outward movement of the ice, but in places effluent glaciers reach the sea and discharge icebergs, so that part of the outflow takes place in the solid form and is not balanced by thawing. Lauge Koch states that the ice fronts in Greenland are about stationary at present.

His proof that on the north end of the island beyond lat. 80° there is bare ground with some plant growth at levels 1,000 meters above the sea, and that summer temperatures

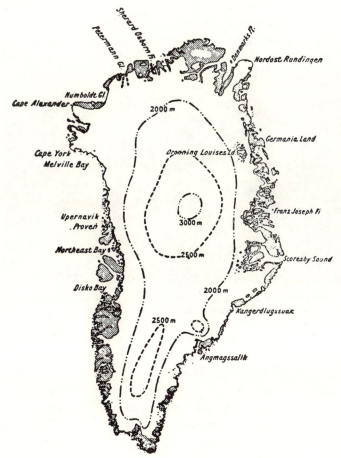

After Lauge Koch.

Map of Greenland.

are higher on the mountains than on the lowlands, is of great interest as showing that elevation in arctic latitudes does not necessarily aid in glaciation.

Greenland probably comes nearer to the conditions which existed during the ice age in North America and northern Europe than does Antarctica, which will be referred to later; and a study of its climate and glacial features enables us to imagine the vast deserts of snow, swept even in summer by blinding blizzards, which covered our own region some thousands of years ago. The real working of the glacial machinery is, however, better studied in the records left in

Inland Ice at Nordenskjöld Glacier, with Nunataks, Sketched from across Billen Bay, Spitzbergen.

the form of drift than in the actual ice field of Greenland, where one sees only the thawing margin or broad glaciers sending icebergs into the sea.

The mountainous rim that encloses most of the Greenland ice sheet prevents its free expansion in many places, and to that extent the great North American and European sheets, which spread out freely on low plains, differed in their mode of operation.

It has been observed by Peary and others in Greenland that there is a dimple or depression of the lofty surface of the inland ice opposite the broader openings in the mountainous barrier, due to the free drainage in those directions; and one may suppose that if on a plain it would expand over a greater area and be thinner in the center than at present.

With so little of its surface free from ice, much evidence of interglacial times is not to be looked for; but from the differing directions of striæ and the greater amount of weathering of Archæan rocks in some places than in others Salisbury suggests that Greenland may have had two distinct glacial epochs.[16]

It has been shown by Peary and others that the Greenland ice dome forms an anticyclonic center from which the air, chilled by contact with the snowy surface, flows off radially in all directions. These outward blowing winds transport much snow to the strip of ice-free land or to the sea beyond, and thus limit the growth in thickness of the ice cap.

REFERENCES

1. A. P. COLEMAN, "The Northeastern Part of Labrador," Mem. 124, *Geol. Sur. Can.*

2. A. P. COLEMAN, "Physiography and Glacial Geology of Gaspé," Mus. Bull. 34, *Geol. Sur. Can.*

3. J. B. TYRRELL, "Compte Rendu," *12th Geol. Congr.;* and *Bur. Mines, Ont.,* Vol. XXII, Part 1, 1912.

4. G. K. GILBERT, *Mon. I, U. S. Geol. Sur.*

5. J. C. RUSSELL, Mon. XI, *U. S. Geol. Sur.*

6. SAMUEL CALVIN, *Bull. Geol. Soc. Am.,* Vol. XX, 1909, pp. 341-356, and Vol. XXII, 1911, p. 207.

7. G. F. KAY, "The Origin of Gumbotil," *Jour. Geol.,* Vol. XXVIII, 1920.

8. Guidebook, No. 6, *Geol. Congress,* 1913, pp. 11-31.

9. *Bur. Mines, Ont.,* Vol. XII, 1904, Part 1, pp. 135-197; and Vol. XX. Part 1, pp. 234-238.

10. *Geol. Sur. Can.,* 1890-91, p. 307 E.

11. W. A. JOHNSTON, *Geol. Sur. Can.,* Mem. 135, pp. 39-47.

12. J. HARLAN BRETZ, *Bull. Geol. Soc. Am.,* Vol. 34, pp. 573, 580 and 602.

13. ARTHUR BEVAN, *Jour. Geol.,* Vol. XXXI, p. 464.

14. A. DE QUERVAIN, *Geogr. Rev.*, 1923, pp. 445-453, "Ice Field of Central Greenland."

15. LAUGE KOCH, *Jour. Geol.*, Vol. XXXI, pp. 42-65, "Some New Features in the Physiography of Greenland."

16. The glacial features of Greenland are well described in various numbers of the *Journal of Geology* during 1894-5, by T. C. CHAMBERLIN and W. D. SALISBURY.

CHAPTER II

Glaciation in Iceland and Spitzbergen

ICELAND and Greenland should interchange names, since Iceland has only 5,500 square miles of glacier, while Greenland is green only along a narrow fringe during a few months of the year. Iceland was much more extensively glaciated, however, during the Pleistocene than now, having been completely ice covered more than once during that time.

Iceland is one of the regions of active volcanoes and it has been desolated in parts by great eruptions within historic times. That conflicts between the contrasting forces of vulcanism and glaciation have occurred in the past is shown by Helgi Pjeturson, who has found four morainic beds separated by volcanic materials. In Burfell, for instance, two sheets of boulder clay are parted by 150 or 200 feet of basalt. The surface of the basalt beneath the upper boulder clay is well striated. There were, then, interglacial periods during which volcanoes were active as in recent times.[1]

It may be, however, that unusual volcanic activity thawed the Pleistocene snows, and that the access of warmth was due to a supply from the earth's interior rather than from an amelioration of climate.

Spitzbergen, lying much farther north than Iceland, is at present much more heavily glaciated, though considerable areas are freed from snow during the brief summer. The inland ice of the northeast island of the archipelago still covers most of its surface, and broad glacial fronts in both of the two largest islands reach the sea. As in the case of the tidewater glaciers of Alaska, there have been striking recessions of some of them within recent years.

That the ice on the southwest islands was once much more extensive than now is evident from the moraines and glaciated surfaces to be seen. Raised beaches show that the land has risen in response to the partial relief from the load of ice.

Glaciation of Northern Europe

The maximum glaciation of northern Europe covered about 2,000,000 square miles, one-half the area of the three great ice sheets of North America. It probably began with much enlarged glaciers in the Scandinavian mountains. The glaciers flowing eastwards expanded on the plains of Sweden and at length the whole of the lowlands with the gulf of Bothnia became a radiant center from which ice moved outwards in all directions. The basin of the North Sea was filled and Norwegian erratics were transported to England and Scotland. In the British Isles the Scandinavian ice joined local mountain glaciers and the whole territory was flooded with ice except the south of England below lat. 52°. To the east most of Russia was occupied and toward the northeast ice extended somewhat beyond the Ural Mountains into Asia. Its most southern points were reached in lat. 50° in Germany and Russia.

The ice sheet was more than 5,000 feet thick over the Bothnian basin, since it was able to cross passes through the Norwegian mountains, some of which reach that elevation.

The drift of northern Europe is divisible into at least two parts, an older, greatly weathered part and a younger fresher one which does not reach so far south. Interglacial

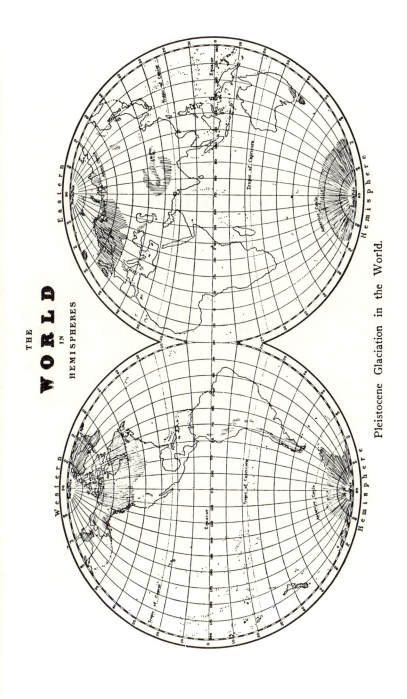

THE
WORLD
IN
HEMISPHERES

Eastern Hemisphere

Western Hemisphere

Pleistocene Glaciation in the World.

beds have been described from England, Germany, Denmark and Russia. That there was a great ice retreat with mild climate as far north as central Denmark is certain, but many English and European geologists do not believe that the ice was completely removed in the time between the deposit of the two boulder clays. The latest account of the plants of the interglacial beds, by Gagel in Germany, shows that the climate was probably milder than at present, especially in the winter, and the finding of hippopotamus, a mammal requiring open water, supports this view. He proves that there was a difference of 10° or 12° Cent. between the temperature of the interglacial time and that of the advances of the ice sheet.[2]

Gagel's conclusions strongly support the idea of a complete removal of the ice at least once in the Pleistocene and bring Europe into line with North America in this respect.

Two or three important ice advances, separated by interglacial times of warmer climate than the present, have been demonstrated in Denmark by the finding of beds of marine as well as of land formation containing remains of both plants and animals belonging to more southern latitudes than the present, so that the Danish geologists reinforce the conclusions reached in Germany.

I have not found references to undoubted interglacial beds in Norway or Sweden, but these countries at the center of glaciation are not likely to afford such proofs, for reasons given earlier; and the same is true of Finland. Interglacial beds of some importance have been reported by Boguloboff in Russia. They include plants and animals indicating steppe conditions.[3]

A very interesting feature of the late glacial history of Sweden and Finland was the formation of great lakes along the ice front, in which annual layers of clay were laid down, "varves," which are connected with a succession of small moraines and have been shown by De Geer to provide an accurate chronology of the retreat and final removal of the ice in the Baltic region.[4]

It is possible that varves of this kind may be found to correspond on the two sides of the Atlantic, thus connecting the series of late glacial events in Europe with those of North America; but thus far De Geer and his former student, Ernst Antevs, who is now studying the seasonally banded glacial clays of North America, have not discovered such a correspondence.

The Ice Age in the Alps

Glacial features are much more completely displayed in mountain regions than on plains, and the glacial history of the Alps has been worked out in great detail by Penck and Brückner. Their report, with its 1199 pages, is the most monumental work devoted to the Pleistocene history of any part of the world; and their results are much more definite and complete than anything published in regard to the great ice sheet of northern Europe.

The Pleistocene record of the Alps begins, of course, with comparatively small mountain glaciers like those of the present day. These extended downwards as the climate grew colder and the snowline was lowered until great valley glaciers moved outward in all directions and expanded over the plains at the foot of the mountains.

The higher mountains were not covered but projected as ridges and isolated peaks or nunataks in the way described in the accounts of the Cordilleran and Greenland ice sheets.

Penck and Brückner have worked out a complicated series of deposits, glacial and fluvio-glacial, and have proved that the snowline was lowered and that the valley glaciers were greatly enlarged four times, with three interglacial times between; as shown in the following table, the Würm being the latest and the Günz the earliest:

Würm glacial period
Interglacial
Riss glacial period
Interglacial
Mindel glacial period
Interglacial
Günz glacial period.

The moraines of the Riss period extend beyond those of the Würm, proving a more powerful glaciation.

The last interglacial period (Riss-Würm) has furnished many fossils, including trees indicating a temperature like that of the present and on the south side of the Alps distinctly warmer. The evidence for the older interglacial climates is naturally not so complete, though walnut, horsechestnut and grape occurring with brown coal in the Mindel-Riss time prove a climate no colder than the present.[5]

The conclusive proof of interglacial periods in the Alps makes it certain that similar breaks occurred also in northern Europe, though the evidence there is less decisive. Changes of climate affecting the Alps so profoundly must have extended over the whole of northern Europe.

The other European mountain regions, the Pyrenees and Apennines, show evidence of Pleistocene glaciation, also, but less is known of the details of their history, and the same is true of the Atlas mountains across the Mediterranean in north Africa.

The Ice Age in Asia

The lowest winter temperatures in the world have been recorded in Siberia, where in places the ground is perpetually frozen for hundreds of feet in depth, and the longest known mountain glaciers exist in Himalayan valleys, yet no evidence of any great continental ice sheet has been found on this largest of the continents, though there were relatively small areas of ice east of the Urals and in part of northeastern Siberia. Probably this is accounted for by the distance of its interior plains from warm seas, and the fact that the winds have to cross ranges of mountains before passing inland, so that the moisture gathered from the ocean is lost on the Himalayas and other lofty chains before reaching Siberia.

There were Pleistocene glaciers, however, on mountains now free from ice, and existing mountain glaciers extended far below their present limits. Lebanon, the Caucasus, the Himalayas and the mountains of Manchuria all show ice-scored surfaces and moraines reaching thousands of feet lower than now, even extending to 4,500 feet above the sea in lat. 26° in Bengal. It is evident that the snowline was lowered proportionally to that of the Alps to the west and the Rocky Mountains to the east.

The absence of great ice sheets in Siberia is paralleled by the unglaciated areas of Alaska and the Yukon Territory in North America, where lofty mountains cut off the supply of snow from the plains inland in spite of a severe winter climate.

One may conclude that the north central plains of Asia largely escaped glaciation because they were sheltered from the moisture-bearing winds, unlike the corresponding plains of Europe and America, which are open toward the Atlantic.

Pleistocene glacial deposits have not been very extensively studied in Asia and, so far as I am aware, only one writer, De Filippi, has mentioned interglacial periods. In the Upper Indus Valley he found four extensions of the ice with three interglacial periods between, and compared them with the glacial and interglacial times of the Alps as described by Penck and Brückner.[6]

The Ice Age in Tropical and South Temperate Lands

Ice sheets on lowlands are, of course, not to be looked for within the tropics, even when the Pleistocene attained its maximum of cold, but wherever tropical mountains rise to the proper elevation glaciers occur upon them at present, and during the ice age these were greatly extended, old moraines usually reaching 3,000 feet or more below the present end of the ice.

The best illustrations of this are found in the Andes, where most of the great peaks show evidence of a greatly lowered snowline.

The loftier volcanoes of Columbia and Ecuador, except a few of the recently active ones, all show evidence of much more extensive glaciation than at present, as noted by almost all climbers who have visited them. The best account of the glacial features of these mountains has been given by Hans Meyer,[7] who describes a number of them which he studied as showing a former extension of the existing glaciers on the average a thousand meters below the present. The old moraines at the lower levels are of two distinct ages, as shown by weathering and erosion, with a long interglacial time between them. Lower down there are proofs of corresponding pluvial periods, as in the Rocky Mountains of the western United States. He correlates the two glacial advances with the Riss and Würm periods of the Alps.

In the Bolivian Andes, which are not volcanic, the same glacial features are found, and near La Paz boulder clay extends below 12,000 feet, evidently deposited by a greatly lengthened glacier coming from Illimani, where, according to Conway, the snowline is now at 16,000 feet and only small glaciers exist.

Moraine at 9,000 Feet, Puente Del Inca, Argentina.

The late Professor Tight, who worked for some time in Bolivia, informed me that he had found evidence of three interglacial periods in deposits near Illimani and Sorata, but that he had published nothing on the subject.

A pluvial period on the tableland of Bolivia and Peru, corresponding to the glaciation of the mountains, is proved by the old beaches rising fifty feet or more above Lake Titicaca, formed when the great desert basin was filled and overflowed near La Paz, forming a parallel with Lake Bonneville.

The only other tropical region, including glaciers at the present day, is central Africa, where Ruwenzori, Kenia and Kilimandjaro display small snow caps and glaciers almost under the equator. Ancient moraines going 3,000 or 4,000 feet below the existing glaciers have been described by a number of travellers, but there appears to be no reference to an interglacial period.

South of the tropics the Andes show very similar relations to those mentioned above, the glaciers having been far longer in the Pleistocene than now, as may be seen near Aconcagua, where present glaciers end between 12,000 and 13,000 feet, while old moraines go down to 9,000 feet along the Transandino Railway.

Except in Antarctica, which will be treated separately, the Southern Hemisphere presented little opportunity for glaciation of the continental type, only South America reaching latitudes where a Pleistocene ice sheet could approach sea level. In Patagonia south of lat. 38° a considerable area was glaciated, as shown by Moreno,[8] Bailley Willis [9] and others. Here, too, there is a division in the sheets of boulder clay with an interglacial period, when, according to Moreno, *Neo Mylodon,* a giant sloth-like animal, still survived.

In South Africa no glacial deposits have been reported,

so far as my reading goes, though one might expect that the highest points of the Drakensberg mountains would have risen above snowline at that time. Passarge, however, has found proofs of alternating pluvial and dry periods in the Kalihari Desert, which may correspond to the glacial and interglacial periods of other regions.[10]

In Australia, David and other geologists describe two periods of glaciation on Mt. Kosiusko and neighboring summits, where at present no glaciers exist.[11]

Tasmania, to the south, was heavily glaciated in the Pleistocene, the ice reaching almost to sea level; and the south island of New Zealand had a great extension of its mountain glaciers, which in some of the fiords reached the sea.[12]

Glaciation in Antarctica

Antarctica, with its 5,000,000 square miles of glacier-covered surface, affords the vastest area of ice known to the geologist, surpassing by far the total of the three great Pleistocene sheets of North America; and there is plenty of evidence that the area has diminished to an important degree since its maximum extension. Striated rock surfaces and high level moraines prove that the ice was once far thicker and covered more of the continent and its adjacent islands than now. It probably encroached much farther upon the sea also in some places than it does at present.

The Antarctic ice sheet is often referred to as illustrating the special features of a continental glacier, such as the Keewatin or Labrador sheets in America, but in reality it differs greatly from them in most respects. The Antarctic ice radiates from a tableland thousands of feet above the sea, while the Keewatin and Labrador sheets, and also the Scandinavian sheet in Europe, had their centers on low ground and spread out over areas of much higher ground.

The Keewatin sheet climbed to 4,500 or 5,000 feet in the foothills of the Rockies; the Labrador sheet, 700 miles from its radiant point, almost reached the summits of the Adirondacks at 5,000 feet; and the Scandinavian sheet crossed the Norwegian mountains of still greater height.

In the detailed and competent account given by Wright and Priestley of the Glaciology of Antarctica the statement is made that, "It would be surprising, in our opinion, if the thickness of the ice sheet exceeded 2,000 feet at any point away from the head of a glacier, and this figure seems to us an outside estimate.[13] It appears, therefore, that the sheet, though of enormous extent, may actually contain less ice than any one of the three great Pleistocene ice sheets of the northern hemisphere. Its outward movement appears to depend upon the slope of the land, like a mountain glacier; and the rate of flow of its effluent glaciers runs from a few feet a year to at most 2.8 feet per day, though the Ross Barrier, which is afloat, advances 4 feet a day.[13]

These rates are very slow as compared with the effluent glaciers of Greenland, some of which move 60 feet a day, and the Greenland ice reaches a central elevation of about 10,000 feet, the same as that of the Antarctic ice, though it has an area of only about 600,000 square miles. If the great island is trough shaped, as seems probable, the thickness of ice in its center may reach 6,000 or 7,000 feet, which is much more in accord with the northern Pleistocene conditions than the 2,000 feet of Antarctica.

The Antarctic ice sheet, instead of being typical, is very exceptional. As described by Wright and Priestly, it is in process of starvation, the meager snowfall (equal to only 7½ inches of rain per annum on the Ross Barrier) hardly replacing the amount removed by ablation and drifting, so that the thickness of ice is probably diminishing. The climate is arid and, as suggested by Scott, a rise of temperature would probably increase the snowfall and so add to the thickness of the ice sheet.

The conditions in this intensely cold and dry continent are, then, widely different from those of the glaciated regions of the north, where a lowering of temperature, probably along with an increase in precipitation, favored the production of ice sheets. It seems even that a time of glaciation in the present temperate regions would mean a time of ice recession in the Antarctic continent. There would be a sort of alternation in the glaciation of the two regions.

A point of great interest for the interpretation of Pleistocene conditions in the glaciated parts of North America and Europe is the evidence of a great anticyclonic area over the south-polar ice sheet. South winds flow outwards in all directions from the highest part of the continent and are deflected by the earth's rotation into southeast winds. This implies that upper return currents bring in moister air from all around to cause the snowfall. As mentioned earlier, this feature is shown also in Greenland, though on a less extensive scale, and Professor Hobbs has argued that an ice sheet doming in the center necessarily produces such an anticyclone,[14] which, once begun, forms a huge air engine extending the ice-covered area in all directions by sweeping the loose, sand-like snow outwards toward or beyond the margin. With some modifications Wright and Priestley support the same view.

No evidence of interglacial periods has been reported in Antarctica, but the uncovered margin of the continent is so small and the study devoted to it by exploring expeditions has been so brief that such a discovery is hardly to be expected. For reasons mentioned above, an interglacial period caused by a general rise of temperature in the world might increase the snowfall and enlarge the ice sheet of Antarctica. Under those conditions no evidence in the form of interglacial deposits containing remains of plants or animals could exist, since the ice sheet never ceased to cover the continent.

In Pre-Pleistocene times, as shown by Wright and Priestley, the continent had a much milder climate and was clothed with vegetation and was probably the center of origin of marsupials and flightless running birds.

The Slope of the Surface and the Thickness of Ice Sheets

Ice is a brittle solid and the idea of its moving like a plastic body under the pull of gravity is not easily grasped. The earlier students of glaciation naturally thought of ice sheets as moving down a slope like mountain glaciers. If an ice sheet spread outwards it was because it centered on mountains or a tableland and the motion was due to the slope of the floor beneath.

It has long been known, of course, that the movements of an ice sheet depend on the inclination of its upper surface and not to any large extent on the topography beneath, though valleys and ridges of rock may direct the motion of its lower layers.

The question of what amount of slope is necessary to set ice in motion is of much interest and has been discussed in works on glacial geology. Some early writers believed that an inclination of at least one degree would be required to cause the outward creep of an ice sheet, and on this basis Croll worked out a thickness of twenty-four miles at the center of the Antarctic ice cap; but concluded that half a degree would be sufficient, giving a central thickness of twelve miles.

How extreme even the smaller estimate was is shown by the fact that the highest level in the interior of the ice sheet.

near the south pole, reaches only about 10,000 feet—less than two miles.

Local ice caps on a small scale are common in arctic regions, and where the underlying surface is not mountainous they take the shape of a shield or carapace, rising somewhat steeply at the edge, but flattening toward the center. Continental sheets have the same general form but greatly broadened and with much gentler slopes. The best-known example is the inland ice cap of Greenland, which has been crossed in several places, giving an idea of its shape. Lauge Koch's small map of northern Greenland shows this well and will be made use of in the following account.[15]

There are two foci of glaciation, a larger one somewhat north of the center of the island, reaching 10,000 feet, and a southern one, which is about a thousand feet lower. J. P. Koch, in 1915, crossed the inland ice from northeast to southwest, near the center of the northern dome, and the contours on the map are largely based on his observations. The southwestern slope from the center may be taken as typical. For the first two hundred kilometers the inclination is about $2\frac{1}{2}$ meters per kilometer; for the next hundred and fifty, $3\frac{1}{3}$ meters per kilometer; for the next sixty kilometers the slope increases to $16\frac{2}{3}$ meters per kilometer; and the final thirty kilometers average $33\frac{1}{3}$ meters per kilometer. There is evidently a curve, very gentle at first but growing steeper as the margin is approached. The actual margin of the inland ice is often too precipitous to be scaled.

Northward from the center the contours are wider apart, indicating a more gentle slope, amounting only to about 2 meters in the kilometer for the first 250 kilometers. Apparently the inland ice spreads out on comparatively low ground in that direction and is not confined by a rim of mountains as it is toward the west.

From the figures just given it appears that a continental ice sheet may spread outwards from a center under the influence of gravity when the surface is inclined only a few feet per mile and has to the eye the appearance of a horizontal plain. Among the older geologists Dana came nearest to the truth in assuming a slope of 10 feet per mile for the

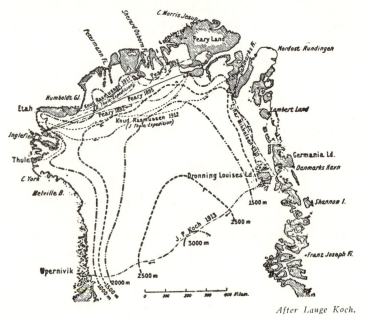

After Lauge Koch.

Map of Northern Greenland, Contours Showing Slope of Ice Cap.

Labrador ice sheet between the northern watershed and the New England mountains.

The thickness of existing ice sheets is not easily estimated and authorities differ greatly in their conclusions. Croll's estimate of the thickness of the antarctic ice, as probably twelve miles at the center, may be contrasted with Wright and Priestley's suggestion that the sheet is not more than 2,000 feet thick as a maximum. As it is known to center on a lofty tableland the latter is much closer to the reality.

The Greenland ice has been estimated by different writers to have a thickness of 2,000 feet or of 6,000 or 7,000. Such estimates depend on one's conception of the shape of the land on which the ice rests. If the island is trough shaped the higher figure may be correct.

Extinct glaciers give more definite answers to the problem. In some cases, as in the Cordilleran ice sheet, there are many nilometers to gauge the flood of ice, since scores of peaks rose above it as nunataks. Erratics, and sometimes boulder clay and striated surfaces, are found in central parts of the region up to 8,000 feet, and several valleys go down to 1,000 or 1,500 feet above sea level; so that at those points there could not have been less than 6,500 or 7,000 feet of ice. The average thickness of the Cordilleran sheet can hardly have been more than 4,000 or 5,000 feet, however.

In regard to the thickness of the Keewatin and Labrador sheets much less is known, since mountains are wanting for hundreds of miles from their centers and the only peaks to serve as gauges are near their edges. There are, however, certain inferences from the results of their work which permit of a rough estimate.

The Keewatin sheet radiated from a center which is not more than 1,000 or 1,500 feet above sea level, and at its southwestern corner it deposited Archæan boulders in the foothills at 4,500 feet, or according to George Dawson, 5,000 feet. Unless the relative elevations have changed since the departure of the ice there must have been, toward the central parts, a thickness of 3,000 or 4,000 feet plus the additional amount necessary to cause a flow for 700 miles or more.

On the southeast the Keewatin ice encountered the

Labrador sheet somewhere in the southern part of Hudson Bay and for a long period of time the two sheets coalesced. Under those conditions the ice reached below present sea level over a part of its area, so that another thousand feet should be added to get the full thickness. This can hardly have been less than 10,000 feet and probably reached 12,000.

Of the Labrador sheet we know the thickness at several points toward the eastern and southeastern sides. It was somewhat less than 2,000 feet thick on the north Atlantic coast of Labrador; not more than 2,000 feet thick in Newfoundland and 2,500 feet thick in the Gaspé mountains, south of the Gulf of St. Lawrence.

In the northeastern United States the evidence is conflicting. Mt. Katahdin in Maine (5,385 feet) is said to have been crossed by the continental glacier, which must in that case have been more than 5,400 feet thick at a point 120 miles southwest of the St. Lawrence at Quebec; and Mt. Washington is said to show glacial material up to 5,800 feet. Some authorities believe the ice crossed over it and must therefore have reached more than 6,290 feet above sea level. Mt. Mansfield in Vermont (4,070 feet) was certainly crossed; and Mts. McIntyre and Marcy in the Adirondacks have glacial materials up to about 4,500 feet, but do not appear to have been glaciated to the summit.

Unless there has been some error in regard to Mts. Katahdin and Washington, the Labrador sheet at points 500 miles southeast of its southernmost gathering ground rose more than 5,800 feet above sea level and probably reached 6,300 feet (Mt. Washington). Allowing ten feet of slope per mile from the center to the southeastern mountains this would mean 11,000 or 12,000 feet for the surface of the sheet east of James Bay. As the Labrador sheet covered most or all of the James Bay basin, this elevation means the actual thickness of the ice. At different times in its his-

tory, according to Low, the glacial center was shifted northward, from lat. 51° to lat. 56°, so that the distance of these
mountains from the center may have been more than 500
miles.

If one omits the Maine and New Hampshire evidence as
uncertain, there remain the proofs that the ice undoubtedly
went over Mt. Mansfield and reached at least 4,500 feet in
the Adirondacks; which can hardly mean less than 10,000
feet at the glacial center when it was at its most southern
point.

Similar evidence is brought forward to show that the
Scandinavian ice sheet was much over 5,000 feet, and probably not less than 10,000 feet thick over the Gulf of Bothnia,
since it crossed the mountains to the west. In Scotland its
margin was about 3,000 feet thick 700 miles or more from
the center, which suggests the same result.

REFERENCES

1. *Scottish Geogr. Mag.,* Vol. XVI, 1900, No. 5; and *Jour. Geol.,*
   1900, p. 282.
2. C. GAGEL, *Zeitschr. d. Deutsch. Geol. Gesell.,* Monatsberichte, Band
   75, 1923, p. 28.
3. BOGOLUBOFF, *Zur Geolog. Geschichte des Gouvernements Kalugo
   in der Glazialperiod,* Moscow, 1905.
4. *"Compte Rendu," Geol. Congr.,* 1910. "A Chronology of the last
   12,000 years," pp. 241-253.
5. *Die Alpen in Eiszeitalter,* Penck and Brückner, 1909.
6. "Review of Himalayan Glaciation," *Geogr. Jour.,* Vol. LXIII, 1924,
   pp. 243-6.
7. *In den Hoch Anden von Equador,* pp. 427-484.
8. *Geogr. Jour.,* Vol. XIV, pp. 368-370.
9. *"Compte Rendu,"* XII, *Geolog. Congr.,* pp. 727-753.
10. *Die Kalihari,* 1904, p. 648.
11. *Austr. Ass. Adv., Sc.,* 1902, p. 204.
12. *"Compte Rendu," Geol. Congr, Mexico,* 1907, pp. 32-3.
13. "British Antarctic Expedition," *Glaciology,* p. 180.
14. *Characteristics of Existing Glaciers,* 1911, pp. 265, etc.
15. *Jour. Geol.,* Vol. XXXI, 1923, Map on p. 49.

CHAPTER III

General Conclusions

FROM the summary just given of glacial conditions during the Pleistocene ice age a number of important conclusions may be drawn which will serve as indications of what must have taken place during earlier glacial periods. The Pleistocene ice age was caused by a lowering of temperature by some degrees over the whole world, since its effects are found in all zones and on all the continents. The production of important ice sheets was, however, dependent on the arrangement of land and sea, on the direction of winds and of ocean currents, and on the relative position of mountain chains, tablelands and plains. The formation of ice sheets was not simply a matter of zones, of distance from the poles or the equator, but was closely related to the present anomalies of climate. The thickest ice sheets were formed on low ground and not on mountains or tablelands, which are the coldest regions at present.

One, or perhaps several, interglacial periods, with temperatures like the present or somewhat warmer, broke the continuity of the glaciation over the whole world with the probable exception of the Antarctic continent.

There was a striking grouping of the greatest ice sheets of the northern hemisphere about the north Atlantic, since 7,000,000 square miles of ice focussed in the lands and shallow seas surrounding it between latitudes 38° and 85°. This is the region where the most important warm ocean current reaches nearest to the pole, suggesting that evaporation from the open sea at points not too distant and not cut off by lofty mountain ranges is necessary for the production of ice sheets.

It is worthy of note, also, in regard to the distribution of glaciation, that Asia, the greatest of the continents within the necessary limits of latitude, bore no important ice sheet, though its mountain glaciers were greatly enlarged, showing that the continent shared in the general lowering of temperature. One may conclude that the central parts of a great land area at a distance from the open sea receive too little precipitation to generate ice sheets on low ground no matter how cold the climate.

Elevation of the land does not necessarily favor the formation of an ice sheet. It may really work against the accumulation of snow and ice. Thibet was not seriously glaciated, but the Baltic region was.

The proximity of a large evaporating surface of sea and an arrangement of the paths of cyclonic storms permitting the transfer of the moisture inland are indispensable for the production of ice sheets of great thickness and activity, such as those of Scandinavia and Canada; since the growth of an ice cap depends on the excess of snowfall over evaporation, thawing, and the outward drift of dry snow from the glacial mass. The arid conditions of the present Antarctic region and the powerful development of the anticyclone over the ice sheet account for its thinness and the relative feebleness of its activity. The Greenland ice cap, in spite of its much smaller size, seems to be more active and efficient, probably because cyclonic storms from the north Atlantic sometimes cross its surface.

Accounts of present glaciation in Antarctica suggest that this largest of ice sheets is much less effective in shaping the continent than the far smaller but better nourished

Pleistocene ice sheets of the Keewatin, Labrador and Scandinavian regions, whose morainic belts often reach hundreds of feet in thickness over a broad territory. It should be mentioned, however, that a considerable part of the drift materials removed by the ice from Antarctica may be deposited out of sight on the floor of the shallow seas surrounding the continent, transported by the vast number of icebergs set free round its shores. Dredging has shown the existence of large numbers of boulders derived from the land on the neighboring sea bottom.

One additional point deserves mention. When glaciers reach the sea they usually discharge icebergs, which float away with the currents until they disappear by melting; but under certain conditions, where the land surface sinks gently under water in a more or less enclosed basin, the ice margin may not break off as icebergs, but may simply encroach more and more extensively on the water area. If the basin is shallow the ice may be thick enough to remain aground until the whole of the area is filled and the water is expelled. This can occur, of course, only in the case of epicontinental seas. The best illustrations are found in the occupation of Hudson Bay by the confluent Keewatin and Labrador ice sheets, and of the North Sea and Irish Sea by the Scandinavian sheet in Europe.

In such instances the whole area of ice may be looked on as continental and the work done is the same as that of ordinary land ice.

There is, however, another possible arrangement, where a somewhat enclosed sea is relatively deep. Then the thickness of the advancing ice sheet may not be great enough to rest on the bottom, so that the expanding edge is afloat but yet remains connected with the land ice. This is the condition in the Ross Sea in Antarctica, where the Ross barrier extends far from land into deep water and rises and falls with the tides.

Such a floating edge of a continental sheet may have considerable additions through snowfall, as has been proved on the Ross barrier, but probably melting by the sea water beneath more than counterbalances the growth from above, so that on the whole the outward moving sheet is growing thinner.

From the floating margin of the Antarctic ice cap great areas are detached without much disturbance, forming the immense tabular bergs characteristic of far southern seas.

This type of discharge appears to have been very important in the southern hemisphere in the Permo-carboniferous glaciation, especially in the Australian region, and will be referred to at greater length in a later chapter.

Effects of Glaciation on Life

Great ice sheets are the most complete deserts in the world, far more desolate than the Sahara or the nitrate region of South America, since almost all land surfaces have an occasional shower permitting a few specially adapted plants to survive, and where there are plants there are almost certain to be animals, though few in species and in numbers.

Life is bound up with liquid water. Water as a gas and water as a solid will not serve the purpose, and in the interior of an ice sheet, away from the edge where thawing may take place in summer and the red snow plant and the glacier flea may subsist, all permanent life becomes impossible.

The formation of an ice sheet means not only that all life is banished from the area covered, but implies also a change in the climate of the adjoining territory, which is chilled by icy winds coming from the broad surface of snow even during summer. It is true that air passing over a high ice dome, like that of Greenland, is warmed in its descent by compression, as in the Föhn and Chinook winds; but this warming is counteracted by the drifting snow particles with which the outward blowing winds are charged and by the heat rendered latent in the melting of ice.

We know that the removal of an ice sheet at the end of a glacial period is a very slow process. In New England, as shown by Antevs from the counting of varves indicating the ice retreat during 4,100 years, it required on the average 22 years for the ice front to recede a mile.[1]

The retreat is often interrupted by long halts or even re-advances. In the withdrawal of the Labrador sheet there was a delay of 7,000 or 8,000 years while the front stood in the Thousand Island region, holding up the waters of Lake Iroquois.

It is evident that the climatic change which causes the removal of ice sheets is very deliberate; and it is reasonable to suppose that the original advance is equally slow and halting, so that there is time for readjustments in the life of the region.

If we begin with the mild climate usual in the history of the world and trace the slow changes due to the coming on of glaciation, we shall expect to find a corresponding adjustment in the organisms of the time, which must either migrate to warmer regions, where this is possible, or become hardened to endure a colder climate on the near approach of the ice front, or perish entirely.

For instance, the warm temperate flora of Dunvegan in Alberta (lat. 56°), which during the Miocene resembled that of the southern United States, grew well within the region afterward covered by the Keewatin ice sheet.

We know that in the Pliocene the climate grew colder and was followed by the glaciation of the Pleistocene.

There are no records showing the life changes near Dunvegan, but in the Klondike, which was not glaciated, though ten degrees farther north, we have evidence of the existence of birch and spruce trees, and of mammoths, horses, bison and deer. The Cenozoic mammals had already adjusted themselves to cold conditions. The stunted spruce and birch were not as thrifty as now and the temperature was probably lower.

While many of the animals changed their habits and their clothing, like the heavily coated mammoth and the woolly rhinoceros of Europe, others simply withdrew toward the south. The hardier forms of plants, too, became acclimated to subarctic or arctic conditions, while the more tender ones perished near the ice but persisted in the warmer, southern parts of Europe and North America.

In interglacial periods the temperate climate flora and fauna followed up the retreating ice front, as shown in the Höttinger interglacial beds of the Alps, and the Toronto Formation in Canada.[2] The return of the ice, after 60,000 or 75,000 years, once more pushed the warm climate forms far to the south. If the classification of the drift adopted in Iowa is correct, these southward and northward migrations took place several times.

In the Pliocene and early Pleistocene, North America was inhabited by a splendid fauna including an assemblage of great mammals like that of Africa at present; but as the succession of glacial and interglacial periods progressed species after species disappeared and toward the close camels, horses, tapirs, mylodons, rhinoceroses, saber-toothed lions, and most of the elephants no longer existed. In the waning stages of the Wisconsin glaciation the last of the mammoths and mastodons and the giant beaver, which had adjusted themselves to earlier vicissitudes, gave up the struggle, for what reason is not known. The musk ox, which had spread to Ontario and the northern states, retreated to the far north, and only a depauperated fauna remained.[3]

In these slow migrations, lasting for tens of thousands of years in each direction, the sifting process was very effective, and at its close the glaciated continent had lost most of its mammals, especially those of large size.

In Europe a similar group of splendid mammals, migrants from Africa at points where the Mediterranean had been bridged, vanished before the end of the period, and the elephant, the rhinoceros, the hippopotamus and the lion were known no more.

In South America the giant edentates, such as the megatherium, and glyptodon, were destroyed, leaving the little sloth and armadillo as degenerate descendants; and in Australia giant marsupials came to an end. The only continents that were little affected were Africa and Southern Asia, where great mammals such as the elephant and rhinoceros and hippopotamus still survive, though in Siberia the increased cold, in spite of the absence of important ice sheets, appears to have killed off the mammoths, mastodons and woolly rhinoceros.

Only two of the land mammals seem to have adjusted themselves completely to the see-saw of cold and warm climates so as to be at home in the tropics as well as the arctic regions, man, who invented fire and clothing, and his satellite, the dog.

Places of Refuge

Some very interesting results of the to and fro migrations of the world's inhabitants are still evident in mountain regions, and one of them, the peopling of high mountains with arctic or subarctic plants, has been referred to frequently. All over North America and Europe above a certain altitude a characteristic flora is found on isolated peaks or ranges, while a very similar one exists on low ground in the Arctic regions. In North America the crowberry, creeping blueberries and cranberries, arctostaphylus, the moss campion (*Silene acaule*). and many other low growing plants occur in this way, often on mountains thousands of miles away from the Arctic circle. For instance they are found above timberline on the Rocky Mountains, on the White Mountains and other parts of the Appalachians, as well as in far northern Canada and Alaska. Similar plants, though usually not the same species, are seen above timberline in the Alps, the Scandinavian mountains and in Arctic Europe.

They all originated in Arctic climates in the far north; but travelled south before the advancing ice sheets to what are now midtemperate latitudes. When the ice receded they were slowly crowded back by the growths adapted to temperate conditions; and also upward toward the mountain tops, where elevation with its lowering of temperature gave them a refuge by halting the hosts of upward creeping temperate forms. The mountains afford islands of Arctic conditions. Similarly isolated mountain colonies of animals occur, but less attention has been given to the fauna than to the flora.

The Life on Nunataks

A very interesting example of the relations of life to ice sheets is presented by the flora and fauna of nunataks, those island-like areas of rock or land completely surrounded by ice. These are not lifeless, as one might expect, but may have a meager plant growth and a few animal inhabitants fitted to the hard conditions.

A small nunatak enclosed in the Brazeau glacier in the Canadian Rocky Mountains was found to have on its rocks mosses and lichens and on scanty patches of soil three

Unglaciated Tableland at 5,000 Feet. Torngats, Labrador.

species of flowering plants, moss campion and two composite flowers, probably dryas. These were visited by flies, which may have flown from lower regions or have been wafted in by the wind.[4]

I have seen no references to the life on nunataks in Greenland, but those of Antarctica seem to be without plants or animals.

Within the last few years the botany of a number of former nunataks, or driftless areas, has been carefully studied by Dr. M. L. Fernald,[5] who finds evidence that the Shickshock mountains of Gaspé, the highlands of Cape Breton Island, and the Long Range of Newfoundland have many endemic plants quite different from those of the surrounding lower regions. It has been proved by the present writer that these are unglaciated areas where the plants seem to have survived all the vicissitudes of the ice age.[6]

The conditions of life on nunataks present singularly dramatic episodes in connection with ice sheets. For instance, as the great Labrador sheet spread eastward filling the lower St. Lawrence Valley, and southeastward over New Brunswick, the highest part of the region, the Shickshock tableland, remained above the sea of white, forming a prison for the plants and animals that inhabited it. The less hardy forms perished, but the more vigorous ones gradually adjusted themselves to more and more severe conditions till the long ordeal was over, a milder climate returned and they began to expand downwards as the ice sheet melted and finally disappeared.

Many of these hardy forms were so changed in the process that Dr. Fernald considers them distinct species or varieties.

The process of elimination of the weaker, less adaptable, forms during ice ages seems to be one of the most effective ways in which new and more viable species arise; so that the ever multiplying plants and animals have their ranks thinned, leaving room for the more progressive species. The hastening and intensifying of the process of evolution in glacial periods is undoubtedly one of the most important modes of developing the life of the world and should receive the special attention of biologists.

Effects of Ice Sheets Combined with Barriers

That all life of plants or animals is banished from the area occupied by an ice sheet is evident, and, as shown already, life must retreat as the ice advances. In North America the way of retreat was broadly open toward the south and southwest, with no transverse ranges of mountains or bodies of water to interfere with it. In Europe east and west mountain chains and the Mediterranean, Black and Caspian seas made a very formidable barrier to the retreat of the inhabitants as the ice advanced and the northern lowlands and the mountains became more heavily glaciated. The barrier was equally effective in interglacial times and at the close of the ice age in preventing the return of many of the fugitive species. This was specially true of the warm temperate forest trees, as has been shown by Berry.[7]

"It is very apparent . . . that the modern species (of trees) with their disconnected distribution represent the segregated remnants of a once world-wide distribution, and that the . . . Ice Age was as unimportant an incident in their history in North America, where there were no mountain or water barriers to cut off their retreat before the ice, as it was a tragic event in Europe, where from Gibraltar to the Caspian a succession of seas and mountains blocked their retreat to the southward." He illustrates this with reference to the walnuts and hickories, as well as other trees, and explains in this way the much greater number of native species of forest trees in America than in Europe.

Unglaciated Surface, a Pleistocene Nunatak. Tabletop Mountain, Gaspé, Quebec.

In the case of Antarctica, once containing ferns and forests and marsupials and running birds, the complete covering of the continent with ice destroyed them all, so that the small fringes of land left free by the retreat of the ice do not now show a single flowering or woody plant nor a land animal above an insect, though the seas are full of vertebrate life, including penguins and seals.

It will be recognized that the complete covering of a land mass with ice annihilates all life, while a partial covering, leaving a sufficiently extensive refuge area, may have comparatively little effect on the plants and animals of the region, all the hardier forms returning, probably even invigorated.

These are points of importance in interpreting the effects of ancient glaciations.

Ice Sheets and Isostasy

It has been proved that in many parts of the world, if not all, the earth's crust is in a state of isostatic equilibrium; that is, that the surface stands at the elevation corresponding to the amount of load in the form of mountains, tablelands or lowlands. It is evident that the piling up of an ice sheet must disturb this equilibrium by adding to the load. This should result in the sinking of the ice covered area to correspond to the increased burden. On the other hand, the removal of the ice by melting at the end of a glacial period should cause a rebound of the earth's crust to its original level. These changes of level should be proportionate to the thickness of the ice deposited and afterwards removed.

In many parts of the world this has been proved by the presence of raised beaches in the glaciated area. Well-known instances are found along the southeastern and eastern border of the Labrador ice sheet.

Fairchild has gathered up much of the evidence and has

Raised Marine Beaches, Spitzbergen.

published a map showing curves of equal elevation circling round a point in northwestern Quebec. The map is founded mainly on the elevation of marine beaches along the St. Lawrence and the Atlantic coast of America, but takes account also of the deformation of the shores of extinct glacial lakes in the interior near the present Great Lakes. Although the curves are in part hypothetical there is no doubt that the map shows a real response of the earth's crust to the relief of load by the removal of the ice.[8]

The highest marine beach which has been certainly determined in the region reaches 690 feet at Kingsmere in the Ottawa Valley. Lower terraces are well seen on Mt. Royal and along the shores of the St. Lawrence, and it has been shown, mainly by Fairchild, that there is a steady lowering as one goes south from the Gulf of St. Lawrence until at New York the old beaches end at present sea level. As New York was the southern limit of glaciation the relations conform to the law of isostasy.

Raised beaches occur all along the Labrador coast and have been found in the Hudson Bay region, so that the area of the Labrador ice sheet has really risen to correspond to the removal of load.

Similar evidence is known on the Pacific coast, showing a rise due to the passing away of the Cordilleran sheet. There is little known as to sea beaches in the Keewatin area; but the deformation of the beaches of ancient Lake Agassis and the old beaches of the Lake Superior region, all of which rise toward the north, prove the same thing in the interior of the continent.

In Europe similar elevation of the glaciated area in Scandinavia,[9] Scotland and England shows that there also there was a response to the relief from load; and there is some evidence of raised beaches in the Arctic islands and Greenland.

The Northern hemisphere conforms to the theory; and probably glaciated areas in the southern hemisphere also rose after the diminution of load, though less is known of raised beaches south of the equator.

It should be remembered, however, that other factors affect the relations of sea and land in important ways during an ice age. The ice itself represents the removal of water by evaporation from the sea, thus diminishing the total volume of the sea and lowering its level in all parts of the world.

The amount of lowering at the climax of the ice age has been variously estimated at from 165 to 500 feet.[10] The restoration of this water to the sea has therefore raised the level everywhere to correspond. It is probable that the ice sheets of Antarctica and Greenland include about one quarter as much water in addition, so that the final sea level, when the whole of the ice has thawed and the temperature of the earth becomes normal again, will be from 50 to 140 feet higher than at present.

It is clear that a great ice age may cause very serious changes in the relative proportions of land and sea, and also that the loading and unloading of glaciated regions may greatly modify the boundaries of a continent, so that the lines of retreat of the inhabitants may be lengthened or cut off during the spread of the icy desert itself.

Duration of the Pleistocene Ice Age

To determine the duration of such an ice age as that of the Pleistocene is not easy and there is much difference of opinion as to the number of thousands of years required for the advances and retreats of the ice and the interglacial times between. The time since the last ice sheet departed has been estimated in several ways, such as the rate of recession of waterfalls, the cutting of cliffs by waves, the building of bars by waves, the deposit of materials in the deltas of rivers, the weathering of glacial deposits, and the counting of annual layers of clay (varves) in glacial lakes.

The last mentioned method, as developed by De Geer and his assistants in Sweden, is the most accurate, giving a reliable chronology for the period covered by the deposits. Up to the present, however, only a portion of the time since the ice began to retreat in Europe has been found recorded in this way; and a still smaller portion has been worked out by Antevs in America.

The retreat of the ice from Southern Sweden to the glacial center at Ragunda (520 miles) took about 5,000 years, and the time since then has been about 8,500, giving a total of 13,500 years since the ice began to leave the region.[11]

The average rate of recession was 1 mile in about 9½ years. The distance between the southern end of Sweden and lat. 51°, in the center of Germany, where the ice terminated, is nearly 300 miles, which at the same rate would require about 2,800 years, the whole time since the ice began to leave Central Europe amounting to 16,300 years on this reckoning. It is very probable, however, that the early stages of the retreat, while the climate was gradually becoming warmer, were very much slower than the later ones when the climatic change was complete. How much should be added on this account is uncertain.

In the Alps, Penck and Brückner, using the rate of formation of the Muota delta in the lake of the Four Forest Cantons, suggest a lapse of from 16,000 to 24,000 years since the Würm glaciation;[12] and they consider that the whole ice age demands several hundred thousand years, the last two interglacial times alone requiring 60,000 and 240,000 years respectively.

In America the length of postglacial time has been estimated in several ways with results varying between 7,000 and more than 39,000 years. The falls of Niagara, which has cut its gorge back nearly seven miles since the ice left the region, has required for the work from 20,000 to 35,000 years, according to F. B. Taylor,[13] and 39,000 according to J. W. Spencer.[14] The present writer, from the cutting back of Scarboro' Heights and the building of Toronto Island by wave work in Lake Ontario, has suggested 25,000 years for the time since Niagara began.[15]

It must be remembered that the ice front had already retreated about a hundred miles before Niagara began its work, so that a considerable addition should be made to these estimates to give the whole time since the maximum glaciation.

From his studies of the varves of New England, Antevs shows that 7,000 years were necessary for the retreat of the Labrador sheet from Long Island to the White Mountains, a disance of 210 miles, making an average of 33 years for a mile. At this rate the time needed for the ice to recede as far as Niagara would be about 3,500 years. Since the beginning of the climatic change was probably gradual the real time may have been much longer.

If Antev's results in New England hold good for the rest of the retreat, the 600 miles between the White Mountains and the glacial center east of Hudson Bay would require 19,800 years, or about 27,000 years for the complete removal of the ice. How long ago this happened there is, at present, no means of estimating. If we take De Geer and Liden's figure of 8,500 years since the ice left Sweden as applying in America, the whole length of time since the Labrador sheet began to wane will be 38,300 years. It seems probable, however, that toward the end melting would go on faster than at first, shortening the time to correspond.

One may conclude that the time since the beginning of the ice retreat in eastern North America can hardly be less than 25,000 years and may reach 35,000 years or more.

The results as suggested above for America seem somewhat greater than those usually accepted in Europe.

If the dissipation of a great ice sheet requires not less than 25,000 years, its formation and advance to points 1,200 miles from its center, as in the southwestern lobe of the Labrador sheet, can hardly demand less time; and it cannot be supposed that as soon as the advance ended retreat began.

It is highly probable that the curve of changing climate was not sharp angled but rounded, so that thousands of years must be allowed for the transition from cold to warm. If we put this rounding of the curve at 10,000 years the whole length of one period of glaciation would be 60,000, or 80,000 years. This estimate should not be looked on as more than a rough guess, though more probably under than over the correct amount in the opinion of the writer.

If five advances and retreats occurred, as suggested in the classification worked out in the Mississippi Valley, this would mean 300,000 or 400,000 years during which ice sheets existed in the Pleistocene.

The length of the interglacial periods, when the climate became mild, is even harder to estimate. Only one of them, that of the Toronto Formation, outlined in an earlier chapter, is well enough known to attempt a time estimate. The following facts are important in such an estimate:

1. A complete change of climate occurred, the temperature at Toronto being several degrees warmer than at present, as shown by the species of trees.

2. The deposits were thick (185 feet), and wide spread, including 80 or 90 feet of thoroughly leached and oxidized clay. The country for at least 64 miles to the north had

been completely weathered, the mud brought down by the interglacial river being free from lime. (The glacial clays of the region are highly charged with lime.)

3. There were great changes of water level amounting to at least 160 feet, and wide and deep valleys were excavated in the delta beds. The best known of these valleys was much more mature than the postglacial valleys in the Toronto region.

The length of this interglacial time has been estimated at not less than 75,000 years. Whether the other interglacial episodes suggested in the Pleistocene of the Mississippi Valley were of similar length is uncertain. If they were, the four interglacial periods would foot up to 300,000 years and the whole length of the Ice Age would be 600,000 to 700,000 years. To the writer this seems a minimum time to allow for all the tremendous and complicated changes known to have occurred in the American Pleistocene.

Chamberlin and Salisbury estimate the length of the Ice Age at from 300,000 to 1,020,000 years, basing their conclusions on the "judgment of five of the glacial geologists who have most studied the data in their most favorable expressions." [16] They conclude their discussion of the subject by the somewhat enigmatic statement: "We place very little value on estimates of this kind, except as means for developing a concrete sense of proportion."

Since their estimate was formed one very important advance has been made giving a more certain foundation for such estimates. The actual counting of years, as developed by De Geer and his assistants in their work on the varves, has given us an accurate chronology of a large part of the last retreat of the ice and has provided reliable determinations of the rate at which ice sheets are melted when the milder climate arrives.

In general, one may say that a great ice age lasts for at

least some hundreds of thousands of years and may continue for a million years or more.

This is, of course, quite insignificant as compared with the greater subdivisions of geological time during which the climate of the world was mild.

The sketch of the salient features of the Pleistocene ice age just given will provide criteria for the interpretation of the features of earlier ice ages; but it must always be remembered that the record of these older glacial periods can never be as complete. Most of the old drift covered surface must always have disappeared by erosion or by depression beneath the sea or must have been buried under later rocks. The record will always be fragmentary and will, of course, become more so as one descends in the geological column, until in the more ancient formations isolated outcrops may be all that remain of what were once vast sheets of till like those of the Pleistocene. In the most ancient tillites, those of the early Pre-cambrian, one must expect to have even these fragments of the history blurred by metamorphism, so that certainty as to their glacial origin may be impossible. Then only the coarser glacial materials, such as moraines and kames formed near the edge of an ice sheet, are likely to be recognizable. Boulder clay will have been transformed into schists, striæ will have disappeared from the pebbles by shearing and slickensides, polished and striated rock surfaces beneath the tillite will almost certainly have vanished through later movements and metamorphism of the rocks.

Our knowledge of the extent and severity of these far off glaciations is dependent on the accident of preservation of remnants of the original tillites in places open to the study of the geologist, and that we have any record of them at all is astonishing.

Of the very early ice ages, in the older Pre-cambrian, we

know and can know very little. It is highly probable that there were important occurrences of glaciation before the Huronian; but final proof of them can hardly be looked for. The instances cited in later pages must be considered as minima and not as giving a full account of ancient refrigerations.

REFERENCES

1. *The Recession of the Last Ice Sheet in New England,* 1922, p. 74.
2. *Guidebook Geol. Congr.,* 1913, No. 6, pp. 13-31.
3. *The Pleistocene of North America and its Vertebrated Animals,* O. P. HAY, Carnegie Inst., Wash., 1923. A very full and excellent account of the Pleistocene changes among American Mammals.
4. *The Canadian Rockies,* "New and Old Trails," 1911, p. 232.
5. *Rhodora,* Vol. XIII, New Series, No. XL, 1911, and later volumes.
6. *Physiography and Glacial Geology of the Gaspé Peninsula,* Geol. Sur. Can., Bull. No. 34, 1922.
7. *Tree Ancestors,* 1923, pp. 78, etc.
8. "Postglacial Uplift of Northeastern North America," *Bull. Geol. Soc. Am.,* Vol. 29, 1918, p. 217.
9. F. NANSEN, *The Strandflat and Isostasy.*
10. DRYGALSKI, *Zeitsch. Geol. Erdkunde,* Berlin, Vol. 22, p. 274; PENCK, *Morphologie der Erdoberfläche,* p. 660, and DALY, *Proc. Am. Acad. Sc.,* Vol. 51, p. 173.
11. DE GEER, *Compte Rendu, Geol, Cong.,* 1910, pp. 241-253; and ANTEVS, *Recession of Last Ice Sheet* in New England, p. XI.
12. *Die Alpen in Eiszeitalter,* p. 1169.
13. Niagara Folio, *U. S. Geol. Sur.,* No. 190, 1913, p. 23.
14. "Evolution of the Falls of Niagara," *Geol. Sur. Can.,* p. 342.
15. "Glacial and Postglacial Lakes in Ont.," *Univ. Toronto Studies, Biol. Series,* 1922, No. 21, pp. 68-70.
16. *Geology,* Vol. III, pp. 413-420.

PART II

PRE-PLEISTOCENE ICE AGES

CHAPTER IV

Probable Tillites in Europe

As compared with the great Pleistocene development of ice sheets there is little to record of ice action in the Cenozoic.

The Pliocene included the cooling down of the earth in preparation for the ice age, and some believe that glaciation began in that period; but most authorities consider that the coming on of the ice sheets marked the transition from the Pliocene to the Pleistocene, so that the whole of the ice age belongs to the latter.

Miocene glaciation on a small scale has been suggested in Iceland,[1] and in the Alps [2] and Apennines; [3] but the general climates of the world seem to have been little affected, and the plant and animal life indicate the usual mildness of nonglacial parts of the earth's history.

An Eocene refrigeration was of much more importance to the world and should receive special mention.

In Europe the late Cretaceous and early Eocene deposits of the Alps include "coarse conglomerates and gigantic erratics of various crystalline rocks. As far east as the neighborhood of Vienna and westward at Bolgen in Bavaria, near Habkeren and in other places blocks of travelled masses appear to have most resemblance, not to any Alpine rocks now visible, but to rocks in southern Bohemia. Their presence has been thought to indicate the existence of gla-

77

ciers in the middle of Europe during some part of the Eocene age." [4]

There has, however, been some discussion as to whether

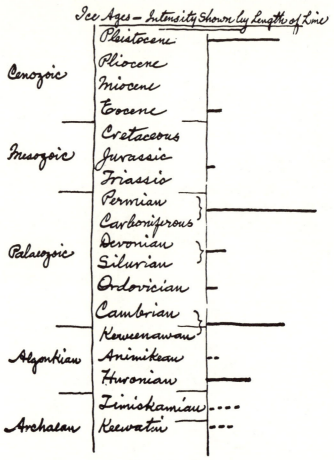

Ice Ages — Intensity shown by Length of Line	
	Pleistocene
	Pliocene
Cenozoic	Miocene
	Eocene
	Cretaceous
Mesozoic	Jurassic
	Triassic
	Permian
	Carboniferous
Palaeozoic	Devonian
	Silurian
	Ordovician
	Cambrian
	Keweenawan
Algonkian	Animikean
	Huronian
	Timiskamian
Archaean	Keewatin

Table of Ice Ages.

these Flysch boulder conglomerates are of glacial origin or not, though it is difficult to account for them in any other way. No striated stones are reported from them, but in a region where mountain-building forces have been so active this is not surprising.

Eocene Glaciation in North America

Eocene glacial deposits are known from several places in the Cordilleran region of North America. The first discovery was reported by C. W. Drysdale in 1911, in the Franklin mining division of British Columbia.[5] His brief account shows the presence of an extensive tillite with many striated stones.

In a visit to the region in 1920, the present writer found that the ancient boulder clay is well exposed for a mile and a quarter in the canyon of a torrent and that blocks of several kinds of rocks, sometimes more than two feet in diameter, are enclosed in an unstratified bed two hundred feet thick. Well striated stones are easily collected and there can be no doubt that the deposit was formed by a glacier.

At present the nearest glaciers are 100 miles away in the Selkirk mountains and end thousands of feet higher up than the Franklin mining camp, which is about 2,900 feet above sea level. The snow line must have been much lower when the tillite was formed.

A somewhat similar boulder conglomerate occurs on the west side of Okanagan Lake, between Penticton and Summerland, in the next valley west of the Franklin region, but no striated stones have been found in it.

A more important area of Eocene glaciation was discovered by W. W. Atwood in 1913 at the base of the San Juan Mountains in southwestern Colorado. It is best displayed near Ridgway, and has therefore been called the Ridgway till.[6] It contains stones of all sizes up to boulders fifteen feet in diameter, and many of them are striated. The till has in places a thickness of 90 feet or more and has been

Ridgway Tillite, Colorado, (Eocene).

Photographed by W. W. Atwood.

followed over a length of eight miles and a breadth of two and a half, covering an area of perhaps twenty square miles. It is thought that the ice must have come from the San Juan Mountains thirty or forty miles away, and as the area where the Eocene till is exposed is flat, it was probably a glacier of the piedmont type.

Photographed by W. W. Atwood.

Striated Stones from Ridgway Till (Eocene).

Atwood's account of this glacial deposit is accompanied by illustrations of striated stones of a characteristic kind; and two specimens sent me by the kindness of K. F. Mather are quite typical. In the summer of 1923 I visited the locality and collected a number of striated stones, and I can entirely confirm the opinion that the deposit is glacial. It has quite the appearance of a Pleistocene till but is much more consolidated.

A boulder conglomerate suggesting a glacial deposit has

been described by G. R. Mansfield from the Eocene Wasatch formation in southeastern Idaho, which may be of the same origin as the two occurrences just referred to, but no striated stones are mentioned, so that the glacial character of the conglomerate is not certain.[7]

During the summer of 1923 the present writer found a tillite east of Gunnison, Colorado, which, in the opinion of R. M. Campbell of the U. S. Geological Survey, is probably Eocene and of the same age as the Ridgway deposits just mentioned.

The tillite rises as steep slopes and pillars to the north of the motor road coming into Gunnison from the east and extends at least three miles, ending, so far as seen, two or three miles from the city. There are boulders up to seven feet in length, and five feet in width, though few reach this size. They include several kinds of granite and of greenstone embedded in a matrix of hardened pale-gray mudstone. The shapes of the stones are typically glacial, but only a few striated stones were found in the hour which could be devoted to the study of the outcrops.

The bed of hardened boulder clay is at least 108 feet thick as measured by aneroid above an irrigation ditch, and it seems to lie horizontal, though no bedding was to be seen.

Later exploration by Professor Atwood and his son has disclosed a much larger area than was seen in my hasty work.

Eocene Glaciation in Antarctica

A moraine-like mass some meters in thickness is mentioned by Wright and Priestley as occurring at Cape Hamilton in Antarctica immediately above Cretaceous beds. "In a clayey matrix lie numerous angular fragments of crystalline rocks foreign to the locality (granite, etc.); also pieces of 'claystone' were noticed. The largest of these lumps of foreign rocks did not exceed half a meter in diameter; most of them were much smaller." The conglomerate may be Cretaceous or early Tertiary, but was certainly formed before the outpouring of the Oligocene basalts.[8]

They quote J. Gunnar Anderson as stating that on the mainland coast between Cape Karl Andreas and Cape Gunnar "A shore nunatak exhibits a coarse conglomerate with boulders up to two meters in diameter. In general the mass is quite unstratified and really much like a bottom moraine, though the rock is old and seems to have taken part in mountain folding." They think that the two deposits may have been more or less contemporaneous.

Ice action at the end of the Cretaceous or in the early Cenozoic has been suggested also in the northwest district of South Australia where large erratics are found resting on Cretaceous beds. It is possible, however, that they have been derived from the weathering of more ancient tillites.[9]

If the Antarctic and Australian boulder conglomerates are Eocene, proofs of glaciation occur on four continents, though none of the tillites are known to have covered a large area. Probably they were formed by local glaciers of a piedmont type, and the refrigeration, as far as known at present, was not sufficient to form ice sheets on low ground. The Eocene cooling did not go far enough to cause an ice age, though it involved an important change of climate as compared with the warm conditions of the early Cretaceous.

Effects on Life of the Eocene Cooling

Although the Eocene refrigeration was comparatively slight it seems to have had an enormous effect on the population of the world. Walther, in his book on the "Geschichte der Erde und des Lebens", devotes a chapter to "Cretaceous time and its great mortality." The sea was cooled and the dominant types of marine life, such as the ammonites and the sea saurians, entirely disappeared at the end of the Cretaceous, while corals and warm-water bivalves diminished in numbers. The clumsy ganoid fishes gave way to the more active bony fishes of our modern seas.[10]

The most striking changes took place on the land, however. The reptiles had reached an astonishing development and had equipped themselves for every kind of life. They ruled the world, so that the Mesozoic is well called the "age of reptiles," and the grotesque and sometimes gigantic dinosaurs, whether herbivorous or carnivorous, formed the strangest and most formidable inhabitants of the land which the world has ever known. During the Cretaceous, the last subdivision of the Mesozoic, when warm temperate forests grew in northwestern Canada, the dinosaurs of the Red Deer Valley in Alberta (lat. 52°) lived and died by the thousand, so that their skeletons are thickly scattered over the "bad lands" along the river. They must have been as numerous as the hardy buffalo of a generation ago, but with much more variety of form and species.

The tiny mammals of the age were apparently few in number and kept out of the way of the ruling vertebrates by burrowing or climbing trees.

The air, too, was dominated by the reptiles, true flying dragons, such as the Pterodactyl and Rhamphorhynchus, and the few birds that existed had teeth and other reptilian features.

Suddenly, at the close of the age, the reptiles of sea and land and air disappeared without leaving a remnant and the coming in of the Eocene, with its cooler, fresher air, rid the world of the monstrous races of reptiles, so that mammals might multiply and grow in stature till they filled the earth in their turn. It was the most dramatic transformation in the history of life on the earth and more than one attempt has been made to account for this surprising turn of the tables between the two classes of land vertebrates.

The suggestion has been made that some fatal epidemic swept the reptiles to destruction, but one fails to see how any single disease could destroy all of them, land saurians, sea saurians and flying saurians at once. It has even been suggested that the little mammals developed the habit of sucking the eggs of the reptiles, thus putting an end to their rule; but this could not exterminate the viviparous marine saurians.

Much more probable is the view that the Eocene fall in temperature, after the long mild ages of the Mesozoic, proved fatal to the cold blooded, unclothed reptiles; while the warm blooded mammals and birds, clothed in their protective coats of fur and feathers, were not greatly inconvenienced and throve mightily when their rivals and oppressors departed.

Thus by a comparatively small change of climate, with cooler winters than reptiles could endure, the life of the world was turned in a wholly new direction, favoring the mammals and culminating in man himself.

But for the slight dip in temperature at the end of the Cretaceous, too small to be called an ice age, who knows but that some elect form of dinosaur might now stand at the head of land animals and some pterodactyl rule the air!

The oncoming of the Pleistocene ice age would probably have halted the career of the cold blooded reptiles, however, giving the oppressed mammals a chance to advance, but the time would have been far too short to allow the wonderful specialization they display at present.

Mesozoic Glaciation in Central Africa

Going back in time from the Eocene few suggestions of glaciation have been found in the Mesozoic, which seems to have been, on the whole, one of the warmest parts of the earth's history. The most striking evidence of ice work comes, strange to say, from Central Africa, where there were great glaciers in the Triassic or Triassic-Jurassic.

The most complete account of glaciation of this age has been given by two American geologists, Ball and Shaler, from the Lualaba Valley west of Lake Tanganyika in lat. 3° 30′ to 5° south.[11]

The tillite forms part of the Lubilache formation, which was deposited mainly in a Triassic lake, perhaps connected with the sea. The glacial beds are partly moraine and partly formed by floating ice. The tillite includes stones of all sizes up to two or three feet in diameter, and the smaller ones, from two to six inches long, are three or four cornered and often striated. The glacial materials came from the south and extend about 100 miles, the northern part consisting of boulders enclosed in shales. It is supposed that the ice had the character of the Malaspina glacier in Alaska.

The evidence of the age of the beds consists mainly of an *Estheria* which Dr. Ulrich considers to belong probably to the Jura-triassic. He believes that it lived in "rather cold water."

The illustrations show excellently striated stones, and the authors think that the glacier may have existed for the greater part of the Permian and Triassic, and add that "when the tropical life flourished in the polar regions a temperate or coolish climate existed to the west of Lake Tanganyika in Central Africa 4° 30′ from the equator."

Three Belgian geologists, G. Passau,[12] P. Fourmarier,[13] and M. Robert,[14] support the views of Ball and Shaler as to the age and character of the tillite, believing it to belong to the Triassic or Permo-triassic and not to the Permo-carboniferous, as one might have expected from the glacial developments in South Africa.

The literature on the subject is reviewed by Hennig, who states that at about the same time evidence was supplied in American, Belgian and German publications, each unaware of the discovery by others of this Triassic glaciation.[15]

The most recent reference to the tillite is by Prof. Asselberghs, who quotes M. Richet as having observed a conglomerate between the Luabu and the village of Mulenga of which "the disproportion of the elements and their form plausibly suggest a glacial origin." [16]

This astonishing find of tillite in Central Africa, almost under the equator, formed at a time when no other continent seems to have been chilled, is very puzzling, particularly if Ball and Shaler's belief that the glaciation took place at sea level is correct. Their suggestion of a great piedmont glacier fed from lofty mountains, since destroyed, seems the least objectionable theory to account for it; though at the present time piedmont glaciers are known only in arctic or cold temperate regions.

One cannot help wondering if these glacial deposits do not really belong to a somewhat earlier time, so as to harmonise with the powerful glaciation on four continents at the end of the Palæozoic.

Jurassic Tillites In North America

The most interesting tillite of the Mesozoic, except those of Central Africa just described, occurs in the Jurassic of California, near Colfax in the foothills of the Sierra Nevada. The breccias of the Mariposa formation of that region have been described by Clarence L. Moody.[17] His map shows the formation as extending nine miles from north to south, with a width of three miles; and a section given includes probably 5,000 feet of interbedded breccia and slate. The breccia beds run from two to four hundred feet in thickness. "All gradations occur in the sediments from fine mud up to masses a yard or more across.   The coarser material is dominantly angular to subangular, although well-rounded pebbles, chiefly quartz, do occur; the matrix is arenaceous to silty.   The recognizable fragments could all have been derived from the associated Calaveras rocks."

Occasionally isolated boulders occur in the slate suggesting floating ice as the transporting agent.

Moody leaves the glacial origin of the breccias in doubt, since he thinks they might have been formed by the work of high-grade and low-grade streams coming from different directions.

A. C. Lawson informs me in a written communication that later study of the field relations shows that the base of the breccias rests on "a smoothly rounded surface of radiolarian chert of Calaveras age (Carboniferous), which surface is strongly scored, fluted and striated, so that I have no doubt but that it is a glacially scored surface and that the breccias are moraines."

The deposits were probably formed by a piedmont glacier.

A deposit probably formed by floating ice has been described by Elliot Blackwelder from the Yakutat region in Alaska as consisting of gritty shale or slate enclosing a variety of pebbles and boulders, often reaching five or ten feet in diameter and sometimes more than fifty feet.   The beds are several hundred feet thick and have been intensely folded, which probably accounts for the fact that no striated stones have been found.   The Yakutat series is probably Jurassic but may be late Carboniferous.[18]

REFERENCES

1. H. C. Ferguson, *Jour. Geol.*, Vol. 14, 1906, p. 122.
2. Hans Schardt, *Bull. Société vaudoise,* Vol. 20, 1884, M. 90, pp. 1-183.
3. L. Mazzuoli, *Com. Geol. Ital.*, 1888, Boll. 2nd, Ser. Vol. 9, pp. 9-30.

4. GEIKIE, *Geology*, 4th Ed., 1903, Vol. II, p. 1239.
5. Mem. 56, *Geol Sur. Can.*, pp. 65 and 95.
6. "Eocene Glacial Deposits in Southwestern Colorado," *U. S. Geol. Sur.*, Prof. Paper 95 B.
7. *Am. Jour. Sc.*, Vol. XLIX, 1920, pp. 401 and 404.
8. *Antarctic Glaciology*, pp. 431-2.
9. T. W. EDGEWORTH DAVID, "Conditions of Climate at Different Epochs," *Mex. Geol. Congr.*, 1907, pp. 29-31.
10. C. SCHUCHERT, "Climates of Geologic Time," *Carnegie Inst., Pub.* No. 192, p. 283.
11. *Jour. Geol.*, Vol. XVIII, 1910, pp. 688-701.
12. "Note sur les depots triassique d'origine glaciare dans la Province orientale (Congo belge)," *An. de la Soc. Geol. de Belgique*, 1912-13, Fas. III, p. 152.
13. Le bassin charbonnier d'age Permo-triassique de la Lukuga," *Ibid.*
14. "Une periode glacière post-permienne dans l'Angola," *Ibid.*, tome XLII, 1918-19, 1°livraison, p. c. 29.
15. "Die Glazialerscheinungen in Aequatorial—und Südafrica," *Geologische Rundschau*, Bd. VI, 1915, pp. 155-165.
16. "Sur la Geologie de la Region Lualaba—Lubudi;" Extr. des *Ann. Soc. Geol. de Belgique*, 1922-3, Liege, 1923, p. 6.
17. "The Breccias of the Mariposa Formation in the Vicinity of Colfax, California." Univ. Cal. Pubs., *Bull. Dep. Geol.*, Vol. 10, No. 21, pp. 383-420, 1917.
18. *Jour. of Geol.*, Vol. 15, 1907, pp. 11, 14.

CHAPTER V

Introduction

THE Palæozoic probably includes a much longer portion of the world's history than the Mesozoic and Cenozoic combined, so that any recurrent phenomenon, such as a cooling of the climate, should manifest itself more frequently than in the times thus far discussed. In reality the Palæozoic shows far more than its proper share of cold periods. Glaciation has been strongly suspected or proved in all of its great subdivisions, the Cambrian, the Ordovician, the Silurian, the Devonian, the Carboniferous and the Permian. The last of the Palæozoic glaciations, which came at the end of the Carboniferous or the beginning of the Permian, is commonly called the Permo-carboniferous ice age; and was much the most severe of the known times of glaciation. All the continents except Europe and North America had large areas covered with ice sheets, which reached within the tropics in two of them and included what are now warm temperate regions in the others. Unless the continents have shifted their positions since that time the Permo-carboniferous glaciation occurred chiefly in what is now the southern temperate zone, and did not touch the arctic regions at all.

In most parts of the world mild conditions returned before the end of the Permian, but, as shown in former pages, the chill seems to have lingered into the Triassic in Central Africa.

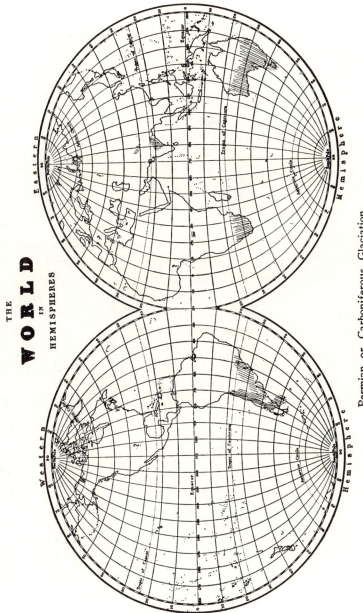

THE
WORLD
IN
HEMISPHERES

Permian or Carboniferous Glaciation.

As this is the most important of the ice ages it will be described in some detail, taking up the continents one by one.

Permian and Carboniferous Glaciation in Europe

Although Europe seems to have been the continent least affected by ice action in the late Palæozoic the first suggestion of a Permian glacier in geological literature comes from the heart of England. In 1855 A. E. Ramsay published a paper "On the occurrence of angular, subangular, polished and striated fragments and boulders in the Permian breccia of Shropshire." [1]

This aroused great interest, since at that time the Nebular Hypothesis as formulated by Laplace was accepted as accounting for the origin of the earth, which was thought to have slowly cooled from a molten state and to be still hot enough at the end of the Palæozoic to give a warm and steamy climate permitting the lavish vegetation of the coal forests.

There was naturally a lively discussion among geologists in regard to the supposed glacial origin of the breccias of the Clent Hills of Shropshire, and the majority were opposed to Ramsay's interpretation. As a rule later British geologists have not considered the Permian breccias glacial, though there is one important exception to this. Oldham, fresh from a study of the Permo-carboniferous tillites of India, believed them to be fluvio-glacial.[2]

In spite of this general doubt as to the correctness of Ramsay's views the Permian breccia has played an important part in the investigation of ancient glaciation, since workers in India, South Africa and Australia were encouraged by Ramsay's paper to study the undoubted tillites in those countries.

Both Ramsay and Oldham refer to the striated stones

found in the breccia, and in spite of the fact that these markings were explained as "slickensides" by other geologists, it seemed to me worth while to visit the region. By great good fortune Mr. Wickam King, who lives at Hagley near Ramsay's localities and who is the best authority on the geology of the district, extended his hospitality to me in 1922 and was good enough to take me in his motor car to the most important localities.

The best exposures of the breccia appear to be at Abberley hill and at Osebury rock, though outcrops occur at a number of other points, the areas being scattered over a length of 75 miles from southwest to northeast with a greatest breadth of 15 miles. They run along the base of an old mountain range, according to Wickam King, and pass in about ten miles into breccia sandstone intercalated with thick marls. The geology has been described by him in several papers, and his conclusions are opposed to a glacial origin of the breccias.[3]

At Abberley hill the breccia is estimated to have a thickness of about 300 feet and more than 60 feet can be seen in one of the pits used for obtaining road metal. The appearance is that of a kame with a vague stratification, and the shapes of the enclosed stones might well be glacial. Many of them are striated but none are typical examples of glacial work.

The blocks are of various kinds, such as Archæan syenite, Cambrian quartzite and fragments of the underlying Silurian.

The Osebury Rock outcrop is similar and a number of striated stones were collected there also.

A careful comparison of the specimens collected with typical glaciated stones from Pleistocene boulder clay convinces me that Wickam King is right in attributing the striæ to internal motions among the pebbles and fragments

enclosed in the breccia. They are slickensided and not glacially striated.

Nevertheless these breccias have the look of glacial deposits. As noted earlier, striated stones are rarely found
in kames, and the absence of true striæ is not final evidence
against a glacio-fluvial origin of the Clent Hill breccias. The
stones are not so sharply angular as they should be if they
are parts of a talus deposit, and on the other hand they are
not so well rounded as they should be if formed by mountain
torrents.

It seems most natural to explain the Permian breccias of
the Clent Hills as formed by local glaciers from the range
of mountains suggested by Wickam King. This, of course,
would imply much loftier mountains than any in Britain or
else a much cooler climate than at present.

On the Continent of Europe glacial deposits have been reported from the Carboniferous or Permian in France and
in Germany. The French occurrence has been described by
A. Julian from the coal measures of central France,[4] where
breccias lie at the base of the formation or are intercalated
in the measures. They present the appearance of moraines
or glacial trains and could not have been formed by torrents
or landslips. The boulders are of all sizes up to twelve or
fifteen meters in diameter and striated stones are found,
though these are exceedingly rare. Good examples of the
breccia are found in the basin of St. Étienne and at Mt.
Grepon, where the deposit is 250 meters thick. The blocks
are of considerable variety, including granite, hornblende
schist and mica schist.

A similar glacial breccia is found at Commentry.

In the same year (1893) a very different type of deposit
was described by E. Kalkowsky from the Frankenwald in
Germany. There is a succession of slates and graywackes,
1,500 or 2,000 meters in thickness, sometimes thinly

laminated, forming layers a centimeter or only a millimeter thick, suggesting a seasonal deposit like the varve clays or slates often associated with glacial beds.   Scattered through this series there are unstratified beds containing pebbles and stones, sometimes only one in a square meter.   The stones vary in size, the largest reaching a diameter of 29 centimeters.   They may be polyhedral, subangular or rounded. No striated ones were found.   The deposits are said to be due to floating ice and not glaciers and the banded shales are apparently looked upon as of marine formation.

Similar stony, unstratified shale has been found in east Thuringia, also.[5]

Scott mentions "undoubted and characteristic glacial moraines, resting upon a polished or striated pavement of Upper Carboniferous rocks in the Lower Rothliegendes of Westphalia (G. Müller, 1901).   They are about four feet thick and suggest local rather than general glaciation.[6]

David quotes Van der Gracht as describing and figuring striated fragments from the Netherlands, but suggests that they are possibly the result of slickensiding.   The same author is quoted as pointing out that striated blocks of hard Carboniferous shale repose on a striated Carboniferous floor at the Preussen Colliery near Lunen; but here, also, there is a possibility that the striæ are due to slickensides since the place is in a zone of faulting.[7]

A supposedly glacial conglomerate is found in the Urals, at the eastern boundary of Russia, in association with the *Gangamopteris* flora.   This will be referred to when the Indian glaciation is described.

Unless European geologists have overlooked evidence of glaciation at the end of the Carboniferous or the beginning of the Permian that continent escaped the worst of the refrigeration which had such overwhelming effects in other parts of the world.   A reason for this exemption is not easily found, but the whole question will be discussed in a later chapter taking up the probable causes of glaciation.

REFERENCES

1. *Quar. Jour. Geol. Soc.* London, Vol. II, pp. 185-205.
2. *Ibid.,* Vol. 50, 1894, pp. 463-8.
3. "Clent Hills Breccia," *Midland Naturalist,* Vol. 16, 1893, p. 25; and also *Quar. Jour. Geol. Soc.,* Vol. 55, 1899, pp. 97-128.
4. *Compte Rendu,* 2nd. Sem., 1893, Vol. CXVII, pp. 255, 46c.
5. "Ueber Geröll-thonschiefer glazialen Ursprungs im Kulm Frankenwaldes," *Zeitsch. d. Geol. Gesell.,* Vol. XLV., pp. 69-86.
6. *Introduction to Geology,* 2nd. Ed., 1915, p. 641.
7. *Trans. Roy. Soc. N. S. W.,* Vol. LIII, 1920, p. 327.

CHAPTER VI

PERMO-CARBONIFEROUS GLACIATION IN INDIA

The Talchirs

FROM Darjeeling and other points in northern India may be seen the magnificent sweep of snowy peaks of the Himalayas, but the snows cease at about 15,000 feet above the sea, and below this level the climate grows more and more sultry until the broad triangle of peninsular India is recognized as one of the hottest regions of the world. Some parts swelter under a moist heat and others parch under desert conditions, but nowhere is there snow or cold even in the winter. It is no wonder, therefore, that the announcement of a glacial deposit in the Talchirs of Central India (lat. 21°) well within the tropics roused incredulity among the geologists of Europe and America. This astonishing discovery was made by the Blandfords while carrying on field work for the Geological Survey of India, and was briefly described and roughly figured in the first report of that Survey in 1859.[1]

They found blocks of granite and gneiss up to four or five feet in diameter in a matrix varying from a coarse sandstone to a fine shale and as they noticed no grooves nor scratches on the boulders they accounted for the deposit as the work of ground ice and not of a true glacier. They were so surprised to find an ice-formed deposit within the tropics that they questioned whether the boulders might not have been carried in the roots of trees, but nothing of a woody nature was found in the deposits, so this idea was given up.

They state that the boulder beds differ greatly from the (Pleistocene) glacial drift or the Permian breccias of England, and hence cannot have been formed by land ice.

The Talchir group of rocks was described as consisting of three parts, a boulder bed at the bottom, followed by a fine

Talchir Tillite, after Engraving in 1st Mem. Indian Survey.

sandstone, on which rests a blue, nodular shale, the whole having a thickness of 500 or 600 feet. From the description given of the shale and from a wood cut in illustration it is probable that this uppermost portion consists of varves. Pebbles and small boulders were found in it with the laminæ of the shale bending round them exactly as in varve clays of the Pleistocene. To this extent there is support for their conclusion that floating ice rather than a glacier did the work; but the unstratified boulder bed at the base must have been made by land ice.

The Blandfords were greatly puzzled at finding bituminous shales, containing coal seams and leaves and stems of ferns (*Vertebraria, Pecopteris, Glossopteris* and *Trizygia*), resting conformably on the glacial beds, since the formation of coal was supposed to imply steamy, tropical conditions. The Damuda beds, as these were named, are the important coal-bearing formation of India.

The economic importance of coal soon led to the study of other regions of the kind, and the same Memoir includes a description of the Rhamghur coal field, about 200 miles north of Talchir, where boulder beds were found alternating with beds of shale and sandstone, the whole having a thickness of 900 or 1,000 feet. This is the first suggestion of interglacial beds in the Talchirs.

There are many accounts of bouldery deposits of this age in central India scattered through later Memoirs and Records of the Survey, but for a number of years little that was new turned up, and some of the geologists of the Survey even cast doubts on their glacial origin, because no striated stones were found in them. Perhaps the most important confirmatory evidence is contained in the third Memoir, where it is mentioned that in the Raniganj coal-field undecomposed feldspar occurs in the sandstone above the tillite, and that there are boulders up to 15 feet in diameter in the lower part of the deposit.[2]

The final evidence that the Talchirs were due to the work of ice was not obtained until 1872, when Fedden discovered striated stones in the tillite and a striated surface beneath it, as noted by Blandford when discussing the geology of Nagpur.[3]

This occurrence was briefly described by Fedden three years later and deserves special mention.[4] The find was made near the village of Irai on Penganga River, in lat. 19° 53', at an elevation of less than 900 feet above the sea.

Typical striated stones were found in the boulder bed and the surface of the Vindhian limestone beneath was scored with striæ for a length of 330 yards. Fedden believed that these striations were due to ground ice and were not the work of a glacier because "there is no commanding elevation of rock older than the Talchirs from which an ice stream could have descended"; and also because boulders had come a long way up the slope from the southwest.

The idea had not yet been grasped that a continental ice sheet can move with little reference to the configuration of the land, because its motions depend on the slope of its upper surface, as shown in a former chapter of this work.

Among later accounts of the Talchirs that of R. D. Oldham in the Geology of India is the most comprehensive and gives a good general description of the tillite and its relationships. One of the most important features brought out is that the tillite rests in cavities of the Archæan rocks, which have the form of *roches moutonnées,* underlying the numerous small coal basins. The similarity of this feature to conditions beneath the Pleistocene till of the Archæan regions of Canada and Scandinavia is manifest.

Oldham describes the glacial beds appropriately as "mud stones" enclosing small and large boulders of various kinds. He mentions that in the Panchet beds, just above the coal, in the upper part of the Lower Gondwana Series, there are bones of primitive reptiles and amphibians; showing that vertebrate life had then returned to the region after the departure of the ice.

The Salt Range Tillites

Long after the discoveries in Central India similar boulder clays were found hundreds of miles away in the Salt Range of the northwest of India. All the ordinary features of glacial deposits are shown in the Salt Range tillites with one

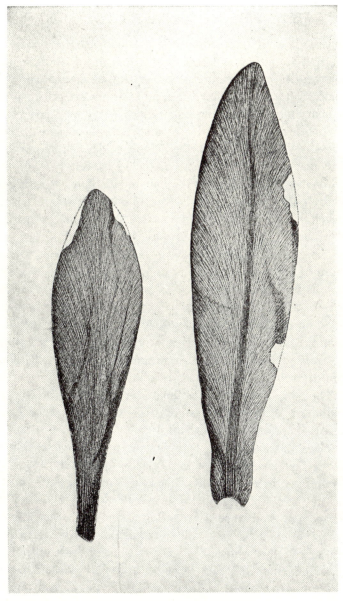

Gangamopteris Major from Karharbari (Talchir) India, from Geology of India.

unusual feature. Sometimes the striated pebbles are sharply facetted as if they had been held firmly in one position till a smooth face was ground and then shifted till another was made. They seem to have been more firmly gripped than in ordinary glaciers, and these sharply carved stones have been compared to the work of a lapidary cutting a gem on the wheel. Was the ice of northern India colder and more solid than that of the great sheets of the Pleistocene?

Some of the boulders of the Salt Range tillite are of rocks whose only known source is 750 miles to the south, which means that the glacier advanced from south to north instead of in the reverse direction, as one would expect.[5]

The inland ice seems to have pushed broadly out to the shore of the northern sea, as suggested by Koken,[6] and to have advanced and retreated, occasionally allowing the deposit of a bed of interglacial sandstone between two beds of till. On the final retreat the Olive sandstone was deposited upon the tillite, perhaps by rivers coming from the retreating ice front. *Conularia* and *Eurydesma* are found at the bottom of the Olive sandstone just above the tillite.

There are the usual cold-climate fern leaves in these beds, and above them, without an apparent break, come the Productus limestones with marine fossils.

The ferns and other plant remains just above the tillites of the Salt Range and of the Talchirs to the south were at first classed by the palæobotanists as Triassic or even Jurassic in their affinities; but the marine fossils suggested a Permian or even Upper Carboniferous age to the palæontologists. For a long time there was a controversy as to the real age of the boulder beds, but in the end the palæontologists carried their point, largely because of the finding of typical marine shells and *Fenestella* of late Carboniferous or early Permian type in the tillites of Australia. The credit for this discovery belongs to Oldham of the Indian Survey.

Those interested in the palæontological discussion will find details given in the works of Waagen,[7] and Koken,[6] since the matter was mainly fought out by Austrian and German investigators. Many references are made to it in the Indian Memoirs and Records, but only an outline of the results need be given here.

Waagen reviewed the whole question from the point of view of the palæontologist, comparing the fossils, both of plants and animals, with those of the corresponding beds in South Africa and Australia as well as those of Europe, and decided in favor of the late Carboniferous age of the tillite. Koken inclined toward a Permian age, while Feistmantel, palæobotanist to the Indian Survey, vigorously supported the Mesozoic character of the beds from the relationships of the plants. Later writers employ the noncommittal term Permo-carboniferous for the late Palæozoic tillites and the overlying coal measures.

To explain the puzzling fact that the plants are much like those of a later age in Europe and North America one may suppose that the glacial conditions in India and the southern hemisphere were too severe for the more tender plants of the coal swamps, which could persist in the milder climate of the northern unglaciated coal regions. Afterwards the hardy southern forms which had developed near the ice sheets spread northwards and were more than a match for the effete Carboniferous plants and gradually replaced them. It required time, however, for their world-wide distribution, so that they reached Europe and North America only in the next geological age.

Many studies have been made of the Talchirs and the Salt Range rocks since the memorable discussion just referred to, but only one need be mentioned here. Recently K. P. Sinor has described a very thin marine bed at the base of the lower Gondwana system (Talchir), in the small

Umaria coal field in the state of Rewah of Central India. *Productus* and *Spiriferina* have been found here. As Umaria is 500 miles from the west coast and 400 from the east, as well as 400 miles from the marine formations of the Salt Range, we have further evidence that the Indian ice sheet spread out over low ground.[8]

Some writers have explained the glaciation as due to the elevation of the gathering ground of neve in Central India; but it is evident that the ice sheet was not of the plateau type but corresponded to the continental type usual in the Pleistocene.

A Visit to the Talchirs

The most interesting locality for the study of the glacial deposits of India is probably the famous spot on Penganga River near the village of Irai in the native states, where Fedden found, for the first time, striated stones and a striated surface beneath the boulder clay, giving indisputable proof of ice work at the end of the Carboniferous. It has, however, seldom been visited since the discovery was made in 1872. The only later account of it which I have seen is that of Sir Edgeworth David, of Australia, who was unfortunate in the time of his visit, since the striated pavement was flooded by heavy monsoon rains.[9]

In November, 1914, through the good offices of the Geological Survey of India, an expedition was made to the place, which is in the Nizam of Hyderabad's dominions in Central India. E. S. Moore and myself were taken in charge by L. L. Fermor, so that the journey was "personally conducted" in the most admirable way.

Crossing an Archæan area with rocks very much like those of the Laurentian and Grenville series of Canada, and afterward the Deccan traps, Chanda was reached, the nearest railway point to the locality. A hot drive in tongas with an

Boating on Penganga River, Central India.

official trotting in front on foot to clear the way, brought us to the village of Irai at the junction of Penganga and Wardha rivers, where a sort of raft consisting of a platform resting on two dug out canoes of teak wood was fastened at the bank of the river.   One of the canoes leaked and the crack was filled with mud to make it temporarily seaworthy.

On this primitive craft, propelled by slender black men with no more clothing than a breechcloth and a turban, it took nearly half an hour to go up-stream a mile and cross the river.

We landed on a broad gently sloping shore with a crumbling cliff capped by a growth of jungle.   The noon sun was very hot in spite of its being the end of November, and yet we stood upon a smoothly glaciated and striated floor of ancient limestone, on which rested about forty feet of typical boulder clay enclosing blocks of granite, gneiss and the local rock, some of them three or four feet in diameter.   Many of them were striated and the crumbling face of the tillite might well have been boulder clay by the shore of a Canadian river.

The striæ on the limestone run, as noted by the early geologists, N. 30° E. to NE., and the motion of the ice was from the southwest.

One can walk on the striated surface for hundreds of yards along the shore and there is no difficulty in collecting striated stones except the heat and perspiration which make the use of the hammer and the carrying of a heavy bag of rocks more toilsome than at home.   The specimens obtained are now in the Royal Ontario Museum in Toronto.

On the way back to Chanda in the evening I walked for a time through bare fields where striated pebbles lay in the soil formed of the crumbling tillite just as in the Pleistocene boulder clay of Ontario.

Our road passed by fields of millet and sessamum, and

Tillite, Penganga River, Central India.

at one or two points went through rank jungle growths, and we saw herds of gazelles and a troupe of monkeys. In one village the water buffaloes were going down to the river to drink, and everywhere the children were clothed only in their own dark skins. Even at the end of November there was no suggestion of cold or of falling leaves; and at Jubbulpore, the nearest city to Chanda, we were told that the thermometer reaches 110° or 115° in the shade in summer.

On a hot evening in early winter two and a half degrees within the torrid zone amid tropical surroundings it was very hard to imagine the region as covered for thousands of years with thousands of feet of ice. The contrast of the present with the past was astounding, and it was easy to see why some of the early geologists fought so long against the idea of glaciation in India at the end of the Carboniferous.

Permo-Carboniferous Deposits in Other Parts of Asia

Outside of India there are comparatively few and unimportant proofs of glaciation, though the depression of temperature must have affected the neighboring regions. Glacial-looking boulder conglomerates with shales and sandstones containing plant remains of the gangamopteris flora have been found by Griesbach at Palezkar near Herat in Afghanistan, five or six hundred miles west of the Salt Range, but no striated stones are reported from them.[10] Some hints of ice work have been found in Kashmir and Thibet also.

Much farther to the west there are boulder beds on the east side of the Ural Mountains, as described by A. Karpinski, who reports that rounded and angular stones up to a meter in diameter are found enclosed in an unstratified matrix, but no striæ have been observed which might not have been caused by tectonic changes, so that their glacial origin is considered uncertain. Cold climate ferns occur in

associated beds, and there are coal seams like those of the
Talchirs at Kusnezk not far away.

The distance between these boulder conglomerates and
the Indian tillites is so great that they can hardly have been
formed by the same ice sheet.[11]

Until the vast interior of Asia has been more thoroughly
explored the real extent of Permo-carboniferous glaciation
in that continent must remain uncertain.

Extent and Character of the Asiatic Glaciation

The Talchir glacial deposits reach lat. 17° 20′ toward the
south and extend northward, according to the geological
map, to lat. 24°, a distance of nearly 500 miles from south
to north. Between this and the Salt Range tillites there are
several hundred miles of later formations, including the
Deccan traps. If we assume that tillite deposits are buried
under these later rocks the length of the glaciated area from
south to north is about 1,100 miles. The map shows lower
Gondwana rocks as extending 600 miles from east to west.

It is understood, of course, that the boulder conglomerate
is not continuous for these dimensions; but it is so widely
distributed that it is reasonable to suppose the whole region
to have been ice covered.

As the Salt Range tillites occur along the foothills of the
Himalayas, which are the loftiest mountains in the world
and bear immense glaciers, one might at first imagine a vast
piedmont glacier at the end of the Carboniferous, spreading
hundreds of miles south of the great range.

This idea is quickly negatived, however, since we know
that the Himalayas are one of the youngest mountain ranges
in the world and were elevated many millions of years after
the tillite was laid down.

There are marine beds with the tillite both in the Salt
Range and at Umaria in Central India, so that the ice

probably reached sea level and was of the continental type, and its direction of motion, as shown by striæ and the transport of boulders, was from the south or southwest, some of the boulders having been carried 750 miles to the north. It did not move from the pole toward the equator.

In most respects the Indian tills are very like those of the Pleistocene, as several of the Indian geologists have suggested. The Archæan boulders so often enclosed in them are just like the granites and gneisses of the Pleistocene boulder clay in Europe and North America; and, except for its consolidation to rock, the matrix is the equivalent of the modern clay. Striated stones are just as frequent and as characteristic as in recent glacial deposits; and in a few places, such as the Ramghur coal field mentioned earlier, there are interglacial beds of shale and sandstone not unlike the stratified clays and sands of the Toronto Formation in America. The thickness of the Palæozoic drift varies greatly but may reach 900 or 1,000 feet, considerably more than any Pleistocene section reported in North America.

The glacial deposits rest on an ancient peneplain of Archæan rocks with gentle mounds and hollows, just as the Pleistocene tills rest on the old Laurentian peneplains of Canada and Scandinavia.

In every respect except one this late Palæozoic glaciation runs parallel to that of the Pleistocene. The known evidence of ice action in India points toward movement only in one direction, northwards, while the Pleistocene sheets spread out in all directions from a center.

Now an ice sheet on low ground, as it seems to have been in India, must necessarily extend in all directions, since it is not the slope of the surface it rests on that sets it in motion, but the thickness of ice toward the central parts, as was shown in the account of the Pleistocene glaciation.

The Indian ice sheet should push southward as well as

northward. Did it really stretch as far to the south of lat. 17° as to the north? It extended 1,100 miles to the Salt Range in the north. If it extended the same distance to the south it would reach the equator!

In the absence of positive evidence such suggestions are, of course, speculative; but the analogy of the great Pleistocene sheets makes it highly probable that the Talchir ice reached much farther to the east and west and south than is known at present.

A similar relationship will be described later in connection with the South African glaciation.

So far as known the Indian ice sheet was the only one of importance in Asia during the Permo-carboniferous. The great areas of the continent to the northwest, north and northeast present few evidences of glaciation, and it would seem as if conditions then were exactly reversed as compared with the Pleistocene, the parts nearest the equator being heavily ice covered while the northern parts escaped.

However, much of what is now northern Asia was then beneath the sea, while India probably reached farther south than at present.

Effects on Life of the Indian Glaciation

India now is a land of luxuriant vegetation, jungle covered in its wilder parts, and birds and mammals swarm wherever man has not conquered the wilderness. The largest land animal living, the elephant, still survives; and this rich fauna and flora exist largely because the Pleistocene ice age had little effect on the climate of the peninsula. Conditions were very different when the severity of the Permo-carboniferous climate with its great ice sheet turned the uncovered fringes of India into something very like the east and west margins of Greenland.

Unfortunately little is known of the land plants and

animals of the Indian region in the Carboniferous; but we may suppose that it was then clothed, at least on the lower ground, with plants like those of the coal swamps of Europe and America, and that gigantic insects flitted among the tree ferns and sigillarias while labyrinthodonts swam in the lagoons or crawled up on the ferny shores.

The ice age at the end of the Carboniferous must have swept all these rank growths away and with them must have vanished the animals that inhabited the swamps. They may have migrated eastwards toward America and westwards toward Europe where no large ice sheets had formed. Those that remained on the lands adjoining the ice sheet adapted themselves to arctic or subartic conditions. Only the hardiest could survive.

The series of rocks beginning with the Talchir boulder clay is called the Gondwana,[12] and this important group has been subdivided as follows:

Upper Gondwana
- Umia and Jabalpore
- Rajmahal and Mahadeva

Lower Gondwana
- Panchet
- Damuda
  - Raniganj
  - Barakar
- Talchir
  - Karharbari
  - Boulder Beds

The stratified parts of the Talchir, or Karharbari beds, above the basal tillite, contain a few plants, mainly species of gangamopteris and glossopteris (*Gangamopteris cyclopteroides, G. angustifolia, Glossopteris communis, Noegerathiopsis hislopi*). These ferns with large, simply outlined fronds were the hardy residue of plants that had developed tissues resistant to cold.

In the time of the earliest coal bearing beds above the

Talchir tillite the climate had grown milder and the number of plants had greatly increased, including not only ferns but equisetums, cycads and conifers.   The only animal remains reported are an insect wing and worm tracks.

In the Damuda beds a labyrinthodont is known, so that amphibians had then returned to the region.   In the Panchet beds there were also reptiles in small numbers (*Dicynodon orientalis,* and a dinosaur, *Epicamopdon indicus*).

In the Rajmahal beds of the upper Gondwana plants become much more numerous and of distinctly Mesozoic type, the commonest species being cycads; while the cold climate ferns, such as glossopteris, were absent.   Evidently the increasing warmth was unfavorable to them.

We may suppose that the cycle of change was then complete and that the normal mild conditions prevailed once more.

Similar changes went on in the sea. As shown by Koken, the waters of the northern ocean close to the ice front contained only a very few hardy forms, such as Conularia and Eurydesma, the former having been found in the boulder-clay itself and both of them occurring in the lowest beds of the Olive sandstone just above the tillite.

Immediately above the Olive formation comes the lower Productus limestone with many brachiopods which could thrive in cool, but not glacial waters.   In the middle Productus limestone nautiloids and reef building corals occur; and in the upper division the ammonites begin and multiply as forerunners of the great Mesozoic development of these cephalopods.[6, 13]   The sea, which had been chilled with floating ice at the broad front of the glacier, gradually warmed up as the ice sheet retreated, and when the ice disappeared soon recovered its usual warmth and was presently swarming with ammonites and belemnites and other life appropriate to the tropics.

REFERENCES

1. *Indian Geol. Sur.,* Mem. I. W. T. and H. F. BLANDFORD, pp. 47, etc.
2. *Ibid.,* Mem III, p. 31.
3. *Ibid.,* Mem. IX, p. 30.
4. *Ibid., Records,* Vol. VIII, 1875, pp. 16-18.
5. *Ibid., Records,* Vol. XXI, Part I, 1888; also *Geology of India,* pp. 120, etc.
6. *Neues Jahrbuch f. Min., Festband,* 1907, pp. 459, etc.
7. "Carbone Eiszeit," *Sitzungsberichte d. K. Boehmische Gesellsch. d. Wissensch, Prag,* S. 102, etc.; and a translation in Records, *Geol. Sur. Ind.,* Vol. XXI, Part 3, 1888.
8. *Nature,* Vol. III, 1923, p. 550.
9. Conditions of Climate, etc., *Mex. Geol. Congr.,* 1907, p. 21.
10. *Ind. Geol. Sur., Records,* Vol. XVIII, 1885, p. 63.
11. *Ibid.,* Vol. XXI, Part 3, 1904, pp. 112-116.
12. *Geology of India,* 1893, pp. 135 and 156.
13. FRITZ NOETLING, *Neues Jahrb.,* Beilageband XIX, 1904, pp. 334-376.

There are references of an incidental kind to Indian glaciation in a number of Memoirs and Records of the Survey beside those given above, which include only the more important accounts.

The Geology of India, both First and Second Editions, treat the subject of Permo-carboniferous glaciation quite fully and should be consulted.

David White gives valuable accounts of the plant remains and their relations, *Am. Geol.,* Vol. III, 1889, pp. 299-330; and also *Jour. Geol.,* Vol. XV, 1907, pp. 615-633.

There is a discussion of the question also in *Lethaea Geognostica,* Theil I, Band II, pp. 579, etc.

CHAPTER VII

Introduction

THE enigmatic glacial deposits of Triassic times a little south of the equator in the Belgian Congo have already been referred to. They may possibly have been formed by a belated ice sheet in a local continuation of the Permo-carboniferous glaciation which was so important in the southern part of the continent.

Boulder conglomerates of a puzzling kind were early found in South Africa, and at first were believed to be of volcanic orgin, "claystone porphyries," etc., but in 1870 they were recognised by Sutherland in Natal as distinctively glacial.[1] He gives an excellent brief description of them as having "a grayish blue argillaceous matrix, containing fragments of granite, gneiss, graphite, quartzite, greenstone and clay slate. These imbedded fragments are of various sizes, from the minute dimensions of sand grains up to vast blocks measuring 6 feet across, and weighing from 5 to 10 tons. . . . The boulder-bearing clay passes into beds which very nearly simulate the condition of true slate but have their lines of cleavage in the direction of, instead of transverse to, the general stratification." Evidently varve shales.

The thickness is in places as much as 1,200 feet, and the old sandstones beneath "in many instances are deeply grooved and striated."

"In some instances ponderous rock fragments are found as much as fifty or sixty miles from the sources of their derivation. . . . The boulder-bearing clay of Natal is of analogous nature to the great Scandinavian drift . . . a vast moraine of olden time."

He refers in support of his conclusion, as the Indian geologists had done, to the Permian glacial deposits described by Ramsay from central England; and in the discussion which followed the reading of his paper before the Geological Society of London, Ramsay himself took part, welcoming the discoveries in Natal as strengthening his own conclusions from the Permian breccias of the Clent hills. It is curious to find the somewhat doubtful and quite untypical English deposits brought forward in support of the glacial origin of characteristic and unquestionable tillites in India and South Africa.

In 1873 a glacial conglomerate was recognised by E. J. Dunn in northern Cape Colony and two years later he gave the name Dwyka conglomerate to southern outcrops of the rock.[2] Since that time the area of known Permo-carboniferous glaciation has been extended to all the provinces of the Union of South Africa and also to the mandated areas of Southwest Africa and Togoland and to southern Madagascar.

The glacial features have proved to be more extensive and wonderful even than those of India, probably implying the most tremendous development of land ice sheets in the world before the Pleistocene. Most of the South African geologists have mapped and described areas of the tillite, Rogers and Du Toit of the South African Geological Survey having been specially active in this field, and the literature on the subject has grown to such large dimensions that much time has been required to collect it, and it cannot be hoped that the list is complete.

An excellent description of the tillite and its relationships is given in the Geology of Cape Colony,[3] where the Dwyka

series is placed as the lowest division of the Karroo system
and as consisting of three parts:

	Feet
Upper Shales	600
Boulder Beds	1,000
Lower Shales	700

The lower shales are absent in most parts of the northern
Dwyka areas where the tillite rests on a glaciated surface of
older rocks.

The latest and most elaborate study of the Dwyka is the
work of Du Toit, "The Carboniferous Glaciation of South
Africa"; [4] and these two works will be largely drawn upon
in the following description. The present writer had an excellent opportunity to study typical sections of the Dwyka
during the South African meeting of the British Association
in 1905, under the able guidance of Mr. Rogers and other
geologists. An interesting account of this visit has been
given by W. M. Davis, who has described what he saw from
the point of view of a geographer.[5]

Dwyka Tillite In Southern Cape Colony

On the 20th. Aug., early spring in the southern hemisphere, our excursion left Matjesfontein in southern Cape
Colony for a tramp of five miles across the Karroo, typical
arid country only sparsely covered with bushes and floored
with gravel or boulders. The low hills of blue gray Dwyka
material with a very rough spiky surface have, at first, no
suggestion of boulder clay. The rock is hard and splintery
and on fresh surfaces looks much like a fine grained basalt,
so that one could understand why the tillite was considered
eruptive by some of the early observers. But presently one
noticed small angular fragments enclosed in it and in some
parts much larger ones, subangular rather than angular, and

of all sizes up to three feet in diameter. They were of varying kinds, such as quartzite, sandstone, granite, and trap or amygdaloid, and had evidently been assembled from different quarters.

Many of the stones which had weathered loose from the matrix were polished and striated on more than one side and some were distinctly facetted. One soon grasped the fact that the rock was a real boulder clay in spite of its hard and brittle character when fresh. Attempts to break out pebbles from the matrix were usually failures, since the fractures passed through matrix and stone impartially, and our specimens were collected on weathered surfaces.

Among the enclosed stones were a few of banded jasper exactly like pebbles in the Huronian conglomerate on which I had worked in northern Ontario a few months before, and in my notes there is an astonished reference to the close resemblance between the two rocks. It may be added that this confirmed a belief which had been growing in my mind that the Huronian conglomerate also was of glacial origin.[6]

An excursion on the following day covered Dwyka ridges a few miles south of Laingsburg, reached by a short railway journey and then by Cape carts, over the same desert surface of stony plains and kopjes of the Karroo as we had seen the previous day but with the interesting addition of the valley of Witteberg River, with a row of forbidding thorn trees along its banks but no water on its stony bed.

Here there was a complete section across the Dwyka beginning with shales without pebbles, evidently a water deposit, followed by well stratified shales containing a few scattered boulders, ice rafted from the front of the advancing glacier in some glacial lake as one may suppose, succeeded without any visible disturbance by typical boulder clay. The glacier had pushed southwards over stratified varve clays

Dwyka Tillite (Squeezed in Mountain Building), Witteberg River, Cape Colony.

without ploughing them up, as ice sheets frequently do when overloaded with englacial materials on their thin outer edge.

About a 1,000 feet of boulder clay, sometimes containing many stones, at others few, are succeeded by 150 or 200 feet of stratified shale, containing only a few small stones, deposited in the lake during a long retreat of the ice. This is followed by the usual unstratified Dwyka, laid down without disturbing the clay. Above this comes "mud rock" with a few pebbles, then shale with no pebbles, but some cherty layers, and the Ecca beds which are partly carbonaceous shales.

The whole series is tilted almost to the vertical and the tillite has a curious wool-sack-like cleavage at high angles to the original structure. The rock splinters into long slender prisms and at first sight has not the slightest resemblance to boulder clay; but striated stones are plentiful among the débris of the weathered surface and there can be no doubt as to the glacial origin of this great series of deposits, which is twice as thick as the greatest development of the Pleistocene complex of tills and interglacial deposits in North America.

Our introduction to the Dwyka glacial beds was about at their most southerly point where, in lat. 33°, they have been caught in the folds of an ancient mountain range running nearly east and west. It may be that they extended for some distance farther south, since the thick upturned edges near Laingsburg are evidently not the original boundary of the formation.

No glaciated floor has been found beneath this southern part of the Dwyka, but a block of quartzite a foot long, polished and striated and deeply grooved on one side and rough on the others, is evidently a fragment plucked from such a floor nearer the center of glaciation. This was found near Matjesfontein and is now, with other Dwyka material, in the collection of the Royal Ontario Museum in Toronto.

Thin sections of the matrix show the usual features of till, fragments of quartz, feldspar and other rock minerals showing angular or subangular forms in finer and finer grains until the final paste consists of particles too small to de-

Striated and Grooved Stone (Part of Floor), Dwyka Tillite, Laingsburg, Cape Colony.

termine, the rock flour of the ancient glacial mill. The matrix for itself may be called most appropriately graywacke, which brings the Dwyka in line with the Huronian boulder-bearing rock, often called graywacke conglomerate.

The Dwyka in Natal

After our first introduction to an undoubted Palæozoic tillite three of the members of Rogers' excursion, Davis, Penck and the writer, decided to drop out of the party and cross South Africa to Natal so as to see other outcrops of this fascinating formation where it rests upon a striated surface of ancient rock.

The very interesting journey across the subcontinent must be left unmentioned; but on the 26th of August, we joined an excursion under the leadership of Messrs. Anderson and

Dwyka Tillite Over Smoothed and Striated Surface, Hlangeni, Natal.

Molengraaff, the latter gentleman meeting us at Vreyheid, whence we drove for thirty miles, passing General Botha's house, destroyed by the British during the Boer War, and following for two miles his fine avenue of eucalyptus trees before reaching after dark our stopping place at Waterval.

Next day a wild drive of a dozen miles brought us to the foot of N'gotche mountain, a flat topped hill capped with a sheet of dolerite and including 2,000, or 3,000 feet of the Ecca-dwyka sediments.

Our special study was devoted to the Dwyka boulder clay in the valley of Hlangeni creek, where it rested on the ancient floor of Barberton slate, Precambrian in age.

The tillite is much like that already described but paler gray in color and of softer texture, weathering down into a close equivalent of Pleistocene boulder clay. Boulders up to five or six feet in diameter occur in the till, many of them from the underlying slate, and some of them are finely striated.

The most interesting feature, of course, was the polished and striated floor beneath, which was as well defined and perfect where just freed from the crumbling Dwyka as any glaciated surface of Archæan rock beneath boulder clay in Canada. There was no room for doubt of the glacial origin of the tillite near Laingsburg; but here in Natal the splendidly preserved surface of rock scoured and carved by the Palæozoic ice sheet was magnificent corroboration of South Africa's great ice age. The Hlangeni creek, though dry at the time of our visit, is evidently re-excavating a Carboniferous ravine and the old walls of the channel now being set free are typically polished and grooved and in places even undercut by the ice like glaciated ravines in Archæan rocks in the Thousand Islands of the St. Lawrence.

Some hours of scrambling and hammering under the intense African sun in lat. 27° 5', without a drop of water, while collecting striated stones and a slab of the polished floor of slate, provided a most impressive contrast between the present and the past, for though Aug. 27 is still early spring, the heat was fully equal to that of a sunny August day in North America. The dry, wilting, sun glare and perspiration made the thought of an ice sheet thousands of feet thick at that very spot most incredible, but most alluring.

The deposits are 480 feet thick and include in the upper parts two thin beds of shale, and still higher a quite thick interglacial sandstone above which there is sandy tillite with many boulders.

The striæ on the Barberton rocks run S 31° E, but the direction of ice motion was clearly influenced by the contours of the land, and the region before glaciation seems to have been of moderate relief, which, of course, was somewhat modified by the ice. The boulder clay was deposited more thickly in the valleys, as suggested by Davis, and the result of the glaciation was to smooth out inequalities.

The whole impression of boulder clays, interglacial beds and striated surfaces near N'gotche mountain was surprisingly like that left on the mind of the student of Pleistocene glacial conditions in the northern United States or Canada, with this strange difference, that the African ice sheet came from the *north*, i.e., from the equatorial regions, while that of America in the Pleistocene was moving toward the equator and not away from it.

This strange fact has been commented on by the African geologists and will receive farther mention on a later page.

After our hot but intensely interesting day it was a special pleasure to accept the hospitality of the nearest Boer farm house, where Mrs. Coetzee gave us, not afternoon tea, but afternoon coffee and delicious oranges from her garden just as the sun was setting behind the mountain. Our wild drive back to Waterval during the night was enlivened by prairie fires through which our Kafir drivers whipped up the mule teams to a gallop. We visited a Kafir Kraal earlier in the day, where half clad negro women and wholly unclad children suggested anything but glacial conditions.

The Dwyka in Transvaal

A visit was made to Vereeniging in the coal-mining region of the southern part of the Transvaal, in lat. 26° 40′, about

200 miles northwest of Vryheid. Here, along the banks of Vaal river, a thin seam of coal rests upon Dwyka conglomerate and is covered by about three feet of coarse conglomerate followed by Ecca sandstones and shales. The Dwyka below the coal partly fills an old ravine in ancient dolomite. It is very bouldery and encloses granites, fragments of dolomite, etc., up to four feet in diameter, many of them well striated. Some of the uppermost stones of the tillite show streaks of carbonaceous matter as if roots of plants had been attached to them. The matrix of the tillite is gritty or sandy rather than clayey.

Some discussion has arisen as to the age of the coal seam mentioned above, certain geologists considering the conglomerate above it a part of the Dwyka, though most authorities believe it to be of later age and not interglacial. The analogy of the important Greta coal beds in New South Wales, which are undoubtedly interglacial, makes the former opinion not impossible, but at the time of our visit no striated stones were found in the upper conglomerate and the latter opinion seemed the more probable.

The Ecca sandstones, which are quarried for building stone two or three miles from Vereeniging, contain in some places crowded leaves of glossopteris and other plants and good impressions can be collected.

There are a number of thick coal seams mined in the region and there is also a curious mine of fire-brick material, where an inclined shaft goes through Ecca shale and sandstone with some coal and penetrates fifty feet into the Dwyka tillite. The matrix of the tillite is made into fire brick and the pebbles and boulders are rejected after passing through the breaker. Striated stones may be found among them without trouble.

Glacial Features near Kimberley

No striated surface was seen beneath the Dwyka at Vereeniging, but a magnificent glaciated surface is found at Riverton near the diamond mining town of Kimberley 250 miles to the southwest. Here again the best exposures are on Vaal River, on low islands or the rocky shore. The glaciated rock is diabase, which shows beautifully rounded *moutonnées* forms. In places a bed of Dwyka covers the glaciated surface but most of it is bare and resembles very much diabase areas under the Pleistocene boulder clay on the north shore of Lake Superior.

Thin coatings of Dwyka may in places be stripped from the polished diabase beneath and show a perfect cast of the striæ. The direction of striation toward the southwest is well seen in many places. In one instance where a block had been plucked from the diabase the cavity was sharp edged toward the northeast and smoothed and rounded toward the southwest.

The polish of the diabase surface has a glistening burnish not seen on similar glaciated surfaces in Canada, probably because a film of iron oxide has been deposited between the till and the diabase. The striæ are visible for a considerable distance away from the weathering edge of the tillite but are gradually lost, though the rounded forms remain.

The smooth black surfaces rising out of the water have tempted the bushmen to artistic work and the forms of animals have been marked on it in places. Many stone tools, arrowheads, knives and scrapers, may be picked up in the region.

A striated surface of similar diabase has been found beneath a thin bed of tillite far below ground in the great open pit of the Kimberley mine; and the old surface under the Dwyka in this region seems to have had only a gentle relief.

Roche Moutonnée, Kimberley, Cape Colony.

Photographed by Young.

much less pronounced than that disclosed in Natal. It distinctly suggests a peneplained surface before the glaciers began their work.

From the evidence obtained during the excursions just described certain points of great interest result. The motion of the ice over the glaciated region was in all cases southerly or southwesterly, instead of northerly as might *a priori* have been expected. The sheet of till is thin toward the north, as might be expected if the glacial center was in that quarter, and thickens toward the south, sometimes reaching 1,400 feet or more. Glaciated rock surfaces are widespread toward the north, but disappear toward the southern edge of the Dwyka, where the tillite passes down into well-stratified shales with no disturbance between.

The points just mentioned are of great significance, as shown by the skilful geologists who planned the excursions and served as guides in the field, and our inferences merely repeat their conclusions.

Extent of South African Glaciation

Having called attention to certain typical areas of the Dwyka it will be desirable now to describe the glaciated region as a whole, as portrayed by the South African geologists and as shown in their maps. In the case of so old a formation it is not to be expected that broad areas of the ancient boulder clay or of the glaciated surface beneath will be available for observation, since undoubtedly many parts have been destroyed by subsequent erosion and many other parts are buried under later formations, but enough is known to show that the greater part if not the whole of British South Africa south of the tropic was ice covered, and that there was glaciation in what was formerly German Southwest Africa and at the southern end of Madagascar.

Du Toit's map shows a width from east to west of at least

900 miles and from north to south of about 700 miles;[4] and he gives reasons to suppose that ice moved westward from land now under the Indian Ocean and perhaps extending to the glaciated southern end of Madagascar.

It is evident then that we are dealing with a glaciated area comparable to one of the Pleistocene ice sheets. He believes that there were three main centers of glaciation but that the ice sheets coalesced when at their greatest development. That the ice moved over great distances is shown by the finding of boulders 800 miles from their sources.

Although he admits that there are some interglacial deposits he believes that these were formed during minor recessions of the ice and that there was no complete removal of the ice as in the case of the Greta coal field of Australia.

In his summing up of the conclusions reached in the carefully reasoned and admirable paper referred to he suggests[4] that "in Pre-dwyka (Lower Carboniferous) times South Africa was a country of low relief, a peneplain in the southwest, rising steadily toward the northeast, where the surface was diversified by low ranges. . . . In the extreme northwest . . . rose another mountainous block.

"The flatter parts of the region could not have stood high above sea level, while a large body of water lay to the south; the higher ground attained altitudes of from 3,000 to possibly 5,000 feet above sea level.

"The general direction of the ice was in a southerly or poleward direction, the northern margin of the compound body, where presumably the motion was equatorwards, having neither been located nor studied yet. In the south the ice probably merged into an unbroken front discharging into standing water.

"The boulder beds represent for the most part ground moraine and north of lat. $32\frac{1}{2}°$ S. rest upon an uneven, polished or striated floor, but to the south of that parallel they pass downwards without a break into deep-water shales that were largely the product of the occasionally buoyed up ice sheet. Just as in North America, the tillite is thin or missing at or about the centers of radiation, while thick away therefrom, sometimes enormously so.

"Oscillations of the ice margins occurred, but truly interglacial periods are as yet doubtful."

Certain features of the South African tillite strike an American geologist as particularly interesting. One is the great thickness of the glacial deposits on the south, reaching, as reported, 1,400 or even 2,000 feet, more than double the thickest known Pleistocene sections in North America. If they were the work of a single ice invasion they must represent the results of a very long period of time, say twice the length of the Pleistocene, or a couple of million years at least; and a vast amount of material must have been transported from the denuded central area. Was the South African ice sheet more effective as an eroding and transporting machine than the Keewatin or Labrador sheets?

Although Du Toit considers the South African glaciation as a unit, unbroken except by relatively small oscillations, I cannot help thinking that in the immense accumulations of drift interglacial materials play a larger part than he allows. Where the Dwyka drift is thick, so far as my own observations go at Laingsburg and N'gotche Mountain, there are always intercalations of stratified clay or sand, now consolidated into shale or sandstone, 50 or 100 or 150 feet thick. Du Toit himself reports an extensive sheet of delta deposits, sandstones and shales, extending for many miles, which implies a great river and a large area uncovered by ice in Natal; and also plant fragments (of *Cordaites* or *Psygmophyllum*) in disturbed sandstone at Matjesfontein, which seem to imply a real interglacial period.

The evidence, as might be expected in a formation pre-

served in so fragmentary a way, is incomplete and unde-
cisive; but would the proofs of the undoubted interglacial
periods with mild climate in the Pleistocene of North
America and the almost equally certain ones in Europe be
any more conclusive if we had only remnants of the beds
disjoined and scattered over the continent?

The great thickness and the frequent occurrence of stratified materials suggest strongly a series of glaciations in
Dwyka times with intervals of milder climate; and the crossing of the directions of striation described by Du Toit and
shown on his map supports this view. The finding by Rogers
of a striated boulder pavement at Eland's Vlei and another
near Riverton point in the same direction.[3]

Another feature which has impressed all geologists who
have paid attention to the Dwyka is the fact that the evidence
up to the present shows only a southward motion. In the
Pleistocene ice sheets there was radiation in all directions,
north, south, east and west, from the glacial center. Why
should the South African glacier not expand northwards as
well as southwards? Can a continental glacier move only
in one direction? One can imagine an ice sheet located on
an inclined plane moving only in the direction of the slope;
but it is worthy of note, as mentioned in an earlier chapter,
that continental ice sheets move out from the center even if
they have to advance up hill.

This was the case with the Keewatin sheet of North
America, which transported materials upwards for 800 miles
with a rise of at least 3,000 feet; and the Scandinavian sheet,
which centered over the Swedish lowlands and pushed across
the mountains which separate Sweden from Norway.

Why then should the South African sheet move in so one-
sided a way?

In considering this question it must be remembered that
South Africa is the only large area of the continent occupied

by white men and explored somewhat carefully by geologists. It is also a singularly open region, largely arid or semi-arid with the rocks splendidly exposed in most places, so that geological exploration is relatively easy; while the rainy and forested region to the north is mostly hidden by dense vegetation and has been studied by only a few geologists in limited areas. Is it not possible that the glaciated area stretched far to the north, perhaps even to the equator?

One recalls that in India an ice sheet belonging to the same glaciation reached at least 17° 20', i.e., 330 miles nearer to the equator than the most northerly point of glaciation reported by the geologists of the Union of South Africa.

In Du Toit's excellent account of the Carboniferous Glaciation of South Africa the northern edge of known ice action is mentioned as 23° S. in Madagascar and 22° in the Union; and as shown on his map, it stretches almost straight across the continent, with the striæ running south or southwest over the whole area. It looks very much as if only the southern half of the glaciated area had been discovered.

The records of Permo-carboniferous glaciation farther north are few and not always very satisfactory. F. P. Mennell reports coal beds underlain "by what is almost certainly the Dwyka conglomerate" in the Tuli district of Rhodesia, somewhere in the neighborhood of Bulowayo.[7] I have not been able to find Tuli on the map, but Bulowayo is in lat. 20°.

In the Belgian Congo, Dwyka conglomerate has been reported several times, J. Cornet apparently having been the first to refer to such a conglomerate, though without certainly fixing its age.[8] O. Stutzer in 1911 described outcrops supposed to be Dwyka at Moshia in Katanga, between 10° and 11° South latitude. Striated stones occur in clay, and a photograph of a quartzite pebble is clearly that of a glaciated stone. The matrix of the tillite was examined by Beck and found to be like that of the Dwyka, and the rock was compared to the Dwyka by Percy Wagner of Johannesburg.[9] In the following year E. Grosse described Dwyka with scratched stones from the Kundelungugebirge.[10]

In 1913 Stutzer described a glacial conglomerate in Katanga, but left the age undetermined since no fossils had been found to settle the matter. The tillite may be older than the Dwyka.[11]

A tillite has been reported by Koert from the Buem formation in Togoland, now a British mandated territory. It is said to occur at Banjeli.[12] I have not been able to consult these works nor to find Banjeli on any map. If the deposit is correctly interpreted the area of Permo-carboniferous glaciation is greatly extended, since the southernmost point of Togoland is several degrees north of the equator.

The facts that glaciation has been proved in Triassic times a little south of the equator and that a glaciation much older than the Dwyka is known from the Congo will make the age of the supposed Dwyka somewhat doubtful until more definite proof is supplied by the finding of associated fossil-bearing beds.

Another point of interest is the altitude of the subcontinent at the time of glaciation. That the preglacial surface was roughly a plain but with considerable local inequalities of hills and ravines is known. Was it near sea level or was it a lofty tableland when glaciation began? Some students of the Pleistocene have assumed that North America was then sufficiently elevated to reach the snowline under ordinary climatic conditions, so that a change of altitude was the cause of the glaciation. Most geologists believe, however, that there was a serious general lowering of the temperature and that ice accumulated on low plains as well as highlands.

It has long been known that the southern edge of the South African sheet reached a large body of water and was preceded and succeeded by water formed shales, as well as having intercalated shales and sandstones. This body of water was lifeless and the seasonal laminations found in places show that it was fresh, since varve clays are not deposited in salt water.

More recently it has been proved in the former German Southwest Africa that the western margin of the ice reached sea level, since marine fossils, *Orthoceras, Eurydesma* and *Conularia,* have been found associated with glacial deposits near Tses and southeast of Keetmanshoop.[13]

It is probable that the whole of South Africa was lower at the end of the Palæozoic than now and it is certain that it was less arid or snow-fields could not have accumulated. Once the process was started the weight of the ice would depress the region. Du Toit suggests a maximum thickness of 4,000 to 5,000 feet at the center; but this seems a very low estimate when one remembers that the Labrador ice sheet not far from its margin nearly reached the tops of the Adirondack mountains, which are 5,000 feet high. Near its center it must have been much thicker, and the same must have been true of the comparable Scandinavian sheet.

Judging from the enormous transport of drift by the South African sheet it seems to have been a much more effective engine than any of the Pleistocene sheets.

It is probable, however, that during the Permo-carboniferous glaciation, when at least four continents were heavily burdened with ice, the sea was greatly lowered, even more so than in the Pleistocene, which would of course broaden the continents and introduce complications which need not be discussed here.

The finding of marine fossils associated with the Dwyka on its western edge brings South Africa in line with the Salt Range in India and with parts of Australia and South America. All of these regions seem to have stood much lower than at present, instead of higher.

Effects of Glaciation on the Life of South Africa

While the ice sheets occupied South Africa all life must have ceased unless some small marginal strips remained uncovered where arctic forms could survive; but of this there is no evidence.

We may think of the present area of South Africa as devoid of life when glaciation was at its maximum. The Dwyka tillite cannot, therefore, be expected to contain fossils unless derived from interglacial or preglacial deposits. The only instance of the sort of which a record has been found is of the plants *Cordaites* or *Psygmophyllum* in fragments of interglacial sandstone from near Matjesfontein.[14]

Yet that the bare boulder clay left by the vanishing ice sheet was soon covered with a hardy vegetation is shown at Vereeniging, where roots of the coal plants pushed down into the till and left carbonaceous traces on the boulders they encountered.

As mentioned on a former page, some geologists think the Vereeniging coal seam to be interglacial, though the majority put it later.

The Karoo system is generally divided into four series, the Dwyka, Ecca, Beaufort and Stormberg series. In most places there seems to be no unconformity between these series and one may consider the upper Dwyka shales, which usually merge into the tillite below, as passing upwards without a break into the Ecca shales and sandstones, and these into the Beaufort beds, etc. The succession of fossils as one ascends from series to series represents, then, the recovery of the organic world from the effects of the ice age.

As might be expected, plants cover the earth before animals arrive and these plants are all cryptogams whose spores could be transported by the wind. Where they came from is uncertain; one would naturally suppose from the unglaciated tropics to the north, but there are suggestions, as mentioned earlier, that the ice sheets may have extended hundreds of miles in that direction to the Congo and Togoland. There may have been a quite different arrangement of land and sea from the present, so that unglaciated refuges for land plants existed not too far off in regions now covered by the sea. Certain theories looking in this direction will be discussed later.

As the Vereeniging coal seams and plant beds come closest to the Dwyka tillite, it may be mentioned that the earliest plant immigrants were ferns, especially the two unfailing genera associated everywhere with the Permo-carboniferous, glossopteris and gangamopteris.

The following list of plants from Vereeniging was made some years ago, *Glossopteris browniana, Gangamopteris cyclopteroides, Noeggerathiopsis hislopi, Sphenopteris,* and *Sigillaria brardi,* all ferns but the last.[15]   Except *Sigillaria* and *Lepidodendron australe,* found elsewhere in Dwyka black shale, ferns seem to have been the hardiest plants of the time, covering the ground with their fronds while the climate still remained cool after the departure of the ice.

The coal seams, which are sometimes twenty feet thick, represent long-continued swamp conditions in a cool climate, like the peat bogs of modern times, and not the tropical conditions generally supposed as necessary for the formation of coal.

In higher levels of the Karroo system more species of plants occur, but the flora is an impoverished one as compared with the luxuriant plant growth of the Carboniferous of Europe and North America.

The known fauna of the Dwyka, aside from a few marine shells and two or three fresh-water forms, consists of a small swimming reptile, Mesosaurus. It was about two feet long and had feeble limbs, but Broom, the main authority on

South African reptiles, believes that it could crawl upon the land, and so was the precursor of the mighty reptiles that ruled the world during the Mesozoic. It is interesting to note that a similar reptile has been found in the Permian of Brazil.[15]

Higher up in the Karroo a few fishes and amphibians are found and reptiles increase greatly in numbers, no less than 73 species being named by Broom in the Beaufort beds. These include herbivores and carnivores, some of considerable size and great specialization, such as the heavily built plant-feeding Pareiasaurus, nine feet long and three and a half feet high, and still larger carnivores which preyed upon them. Some of these reptiles were very like mammals in their bony structure, suggesting that the origin of the now dominant vertebrates may have been connected with this sudden outburst of land reptiles in South Africa after the passing away of the ice sheets which had made a clean sweep of all previous land life.

Whence came these ancestors of the land vertebrates remains a mystery, and just why South Africa, the most heavily glaciated of all the continents, should have been so rich in these strange land reptiles, the precursors of the great hosts which were to follow in other parts of the world during the Mesozoic, is unknown. The other regions furnishing Permian reptilian faunas of considerable interest are eastern Russia and Kansas, which were not glaciated. India also has provided a few. That similar reptiles should occur at the same time in such distant and apparently unrelated countries is astonishing, and can only be accounted for by the loss of many connecting links.

It may be that South Africa was nearest to the point of origin of the land inhabiting reptiles, perhaps in an unglaciated Antarctica, and so was the first to receive vertebrates inhabiting the land.

REFERENCES

1. "Notes on an Ancient Boulder Clay of Natal," *Quar. Jour. Geol. Soc.,* Vol. XXVI, pp. 514-17.
2. *Ibid.,* Vol. 61, 1905, *Glacial (Dwyka) Congl. in the Transvaal,* pp. 473-7.
3. ROGERS and DU TOIT, *Geol. of Cape Colony,* pp. 166, etc.
4. "Carboniferous Glaciation of S. Africa," *Trans. Geol. Soc. S. Africa,* Vol. XXIV, pp. 188-227.
5. *Bull. Geol. Soc. Am.,* Vol. 17, pp. 400-420.
6. *Am. Jour. Sc.,* Vol. XXIII, 1907, pp. 187-192.
7. *Science in S. Africa,* 1905, "Geology of Rhodesia," p. 302.
8. *Observations sur les terrains ancien du Katanga,* Liege, 1897, p. 52 ff.
9. *Zeitschr. d. Deutsch. Geol. Gesell.,* Bd. 63, 1911, Monatsbericht No. 12, p. 626.
10. *Ibid.,* Bd. 64, 1912, Monatsbericht, B. p. 320.
11. *Ibid.,* Bd. 65, 1913, Monatsbericht No. 2, p. 114.
12. *Mittheilungen aus d. Deutschen Schützgebieten,* 1906, pp. 113-131, u. *Erläuterungen zur geologische Karte von Togo.*
13. *Geol. Sur. S. Af.,* Mem. 7, 1916, p. 65.
14. *Trans. Geol. Soc. S. Af.,* Vol. XXIV, 1921, p. 214.
15. *Science in S. Af.,* p. 398.

Many other publications refer to the Dwyka; such as—*Dunn's Report Presented to Parliament,* Cape Town, 1886; SCHENCK, *Ueber Glazialerscheinungen* in S. Af., *Habilitationsschrift, Friedrichs Univ.,* Halle; MOLENGRAAFF, "Glacial Origin of the Dwyka," *Trans. Geol. Soc. S. Af.,* Vol. IV, pp. 103, etc.; ANDERSON, *1st. Rep. Geol. Sur. Natal,* 1902, pp. 14-17; CORSTORPHINE, "A Former Ice Age in S. Af.," *Scot. Geogr. Mag.,* Vol. 17, pp. 57-74; SCHWARTZ, "The Three Palaeozoic Ice Ages of S. Af.," *Journ. Geol.,* Vol. XIV; SEWARD, "Fossil Plants from S. Af.," *Geol. Mag.,* Vol. IV, pp. 481-7; COLEMAN, *Bull. Geol. Soc. Am.,* Vol. 19, 1908; HATCH and CORSTORPHINE; *Geol. of S. Af.,* 2nd Ed., pp. 219-243 and 335-339; etc.

CHAPTER V,III

General Features

DURING the Pleistocene Australia escaped glaciation except on a few of its highest mountains, though the island of Tasmania just to the south was largely covered by an ice cap. Toward the end of the Carboniferous, however, it was widely glaciated and in some localities there is good evidence that ice ages recurred several times with long interglacial periods between. The succession of glacial and interglacial periods probably lasted longer in the island continent than in any of the others, though South Africa excels it in the massiveness and thickness of its boulder clays and also in the area of striated surface beneath.

The first discovery of glaciation in Australia was made by A. R. C. Selwyn in the Inman Valley of South Australia in 1859. He noted its resemblance to glacial features in Wales, but made no suggestion as to the age of the glaciation.[1]

In 1866 R. Daintree reported from Bacchus Marsh in Victoria "Here I have found a few pebbles grooved in the manner I have read of as caused by glacial action";[2] but the age of the beds was not settled till later by E. J. Dunn, who proved that they were Permo-carboniferous.[3]

Similar tillites were found in other parts of Australia and it was presently shown that all the states of the Commonwealth had been more or less glaciated and that in New South Wales and West Australia the ice had reached sea level.

Most of the Australian geologists have paid some attention to the ancient tills and for a number of years a committee of the Australian Association for the Advancement of Science has reported upon glacial matters, so that a large literature has grown up in regard to it. Professors David [4] and Howchin [5] have been particularly active in this respect and their writings include references to the early work. In 1886 Oldham of the Indian Survey visited the New South Wales coal region near Newcastle and proved that the glacial beds known to exist there were connected with marine fossils. He found "blocks of slate, quartzite and crystalline rocks, for the most part subangular, scattered through a matrix of fine sand and shale, and these latter beds contain delicate *Fenestellæ* and bivalve shells with the valves united, showing that they had lived, died and been tranquilly preserved where they are now found and proving, as conclusively as the matrix in which they are preserved, that they could never have been exposed to any current of sufficient force and rapidity to transport the blocks of stone now found lying side by side with them. These included fragments of rock are of all sizes, from a few inches to several feet in diameter. . . .

"It is impossible to account for these features except by the action of ice floating in large masses, and I had the good fortune to discover, in the railway cutting near Branxton, a fragment beautifully smoothed and striated in the manner characteristic of glacier action. . . . This seems to show that the ice was of the nature of icebergs broken off from a glacier which descended to sea level."

It is almost certain that the lowest glacial horizons are

considerably below the top of the Carboniferous and therefore that glaciation began earlier in Australia than in any
of the other continents.   It has been suggested that the
maxium glaciation occurred at different times in the different
continents.   This subject will be discussed later.

Though the first discovery of Permo-carboniferous glaciation was made in South Australia, where the Inman Valley
and Hallet's Cove give unsurpassed examples of tillite and
striated rock surfaces, as described by David, Howchin and
others,[5, 7] the Bacchus Marsh area in Victoria attracted
attention soon after and has been described by even more
observers, including E. J. Dunn, as mentioned before,
Graham Officer and L. Balfour,[8] Sweet and Brittelbank [9]
and numerous others.   It displays a wonderful succession
of tillites and interglacial beds indicating a long succession
of advances and retreats of an ice sheet coming from the
south.

Much has been written also of the glacial and interglacial
beds of the Newcastle coal region in New South Wales,
where Oldham first proved the age of the glaciation by
means of marine fossils, as previously mentioned.   Of late
years this area has become the most interesting of all, since
the beginning of glaciation has been proved to have occurred
in the Kuttung Series of Middle and Late Carboniferous
time, far below the Permo-carboniferous Lower Marine
Series which corresponds to the time of glaciation in India
and South Africa.

Three great periods of glaciation have been demonstrated
here, with two very long interglacial periods, the whole
series of beds having the enormous thickness of 9,800 feet,
including marine deposits and also the Greta Coal Measures
which are of great economic importance.   These are the
most extensive interglacial formations known in the world

and must have required hundreds of thousands or even millions of years for their deposit.[10, 11, 12]

Another very interesting development in the region has been the finding of extensive sheets of varve slates, which will be described later.[13]

The working out of the relations of ice sheets, floating ice, land formations, and coal swamps in the Kuttung and Hunter River districts of New South Wales has introduced new and extremely interesting and important problems in regard to glaciation in late Palæozoic times; and the theories proposed to account for ice ages will have to be adjusted to them.

In addition to the evidence of ice work in the older and more populous states of Australia, glaciation on a large scale has been reported from the island of Tasmania,[14] where E. J. Dunn, T. B. Moore,[15] David, Feistmantel[16] and others have described tillites, interglacial beds and striated surfaces like those just referred to on the mainland. At Wynyard, for instance, 1,220 feet of glacial and interglacial beds are noted by David, with coal measures equivalent to the Greta beds of New South Wales.[17]

On the east coast of Australia, in Queensland, R. L. Jack has reported ice transported blocks in the Bowen River coal area (lat. 21°).[18] Spencer and Byrne have briefly described similar boulder deposits with undoubted glaciated pebbles from Yellow Cliff in the Northern Territory;[19] and on the west coast A. Gibb Maitland and W. D. Campbell have found marine glacial beds in the Gascoyne and Irwin River areas, where they extend for 450 miles along the shore and reach the tropics.[20]

A Glimpse of the Australian Glacial Beds

During the Australian meeting of the British Association for the Advancement of Science in 1914 an opportunity was

Photographed by John Greentees.

Permo-Carboniferous Erratic, Hallet's Cove, South Australia.

afforded to visit glaciated regions in South Australia, Victoria and New South Wales under the best of guidance and with a number of other geologists interested in glacial matters.

In South Australia Professor Howchin led a party south from Adelaide to Hallett's cove on the shore of St. Vincent Gulf where the crumbling surface of tillite closely imitated Pleistocene boulder clay until one attempted to remove stones from the matrix. In parts of the tillite the enclosed pebbles and larger stones were only sparsely scattered through the matrix, as in till at Scarboro on Lake Ontario, but the proportion that were striated was greater than in Canada.

In the Inman Valley the glacial features were still more impressive and here and there one encountered huge granite erratics standing up from the vegetation like perched blocks. They were originally enclosed in the boulder clay, but the easily weathering matrix has been removed leaving them as scattered boulders twice the height of a man.

The underlying floor of quartzite is in some places so polished that the sunlight is reflected, and also finely striated, the groovings indicating a motion of the ice from the southeast. I have never seen a finer example of a glaciated surface in the Pleistocene of North America.

On this excursion tillite was seen from point to point for a distance of thirty-five miles from north to south, with a width in the southern portion of about fifteen miles, and it has been mapped for a number of miles farther to the east.

In Victoria an excursion was made to the famous Bacchus Marsh where we saw typical till with striated stones and a *moutonnée* surface beneath, and also undoubted interglacial beds of sandstone containing fronds of gangamopteris and other plants, evidence of a real withdrawal of the ice suffi-

Polished and Striated Surface under Tillite, Inman Valley, South Australia.

ciently long and mild to permit the land to be covered with vegetation. Previous work by David and other geologists has shown that there are nine or ten glacial horizons with more or less stratified material between, the whole series having a thickness of 1,635 feet.

In New South Wales a party of geologists under the guidance of Sir Edgeworth David and Mr. Süssmilch had an opportunity to see the extraordinarily interesting glacial and interglacial beds in the Newcastle coal region, including the most striking example of a long and relatively warm interglacial period in the world.

At West Maitland we saw the Lower Marine beds containing boulders and pebbles, sometimes striated, and shells of aviculopecten, productus, etc., as well as fenestella, evidently living there while ice floated in the sea and dropped its burden as it thawed. It was in this region that the age of the tillite was first determined by Oldham.

At Branxton there appeared to me to be thin beds of true tillite interstratified with the marine deposits, though the Australian geologists do not interpret any of the beds as due to land ice.

Here we had an opportunity to see the Greta coal beds with tree trunks reaching up into the strata above. There are two seams with a total thickness of forty feet of coal and the enclosing sandstones and shales contain a rich flora including *Gangamopteris, Glossopteris, Sphenopteris, Noeggerathiopsis* and *Dadoxylon*.[11, 12]

Half a mile southwest of Seaham the lowest glacial horizon was found at the time of our visit by David, who places it much below the Permo-carboniferous, perhaps in middle Carboniferous time, putting the commencement of glaciation much earlier than in India or South Africa.

In general the glacial deposits in the Maitland and Hunter River region impressed me often as more like the work of

Glossopteris linearis (right and left), Glossopteris Browniana (middle), from Süssmilch's Geology of New South Wales.

land ice than of floating ice. Much of the matrix showed no stratification, and the boulders of granite, Laurentian-looking gneiss, and rhyolite from a nearby source, were scattered through it as in true boulder clay. One boulder of the latter rock had a diameter of at least 12 feet and could hardly have been transported by floe ice.

It has not been a surprise, therefore, to learn that proofs of the action of a glacier have recently been found in the Kuttung series, the Carboniferous lower portion of the glacial beds of the Maitland District. Osborne and Browne report a smoothed and striated surface from the parish of Wolfingham not far from the region visited by the British Association party.[21] This greatly extends the area covered by the Australian ice sheets, and makes the Greta coal beds still more important as indicating a great warming up of the climate in the interglacial time.

Since our visit (in 1914) another feature of much interest has been described by Süssmilch, David and others, throwing light on the glacial conditions of the time. In many places banded shales or slates have been found, ancient varve clays, like those described by Sayles from the Boston region. These occur from point to point for 200 miles, and in places have a thickness of 200 feet. As the laminæ average two-thirds of an inch across, this implies more than 3,000 years for the deposit.

As in other stratified glacial clays, stones are sometimes embedded in them, dropped from floating ice, and frequently beds of the shale are crumpled from the grounding of ice fields or perhaps of bergs. These finely stratified clays imply marked changes of seasons with rapid thawing of the not far distant ice front in summer and the slow deposit of the finest mud during winter. The varve clays must have been laid down in fresh-water lakes like the glacial lakes in the St. Lawrence basins toward the end of the Pleistocene ice age.[12] An excellent description of these features, accompanied by good illustrations, has been given by Süssmilch in the report previously referred to.

Special Features of the Late Palæozoic Glaciation in Australia

In certain ways the Australian glaciation differs greatly from that of Permo-carboniferous times in the other continents. It seems to have begun much earlier than in India and South Africa, but, judging by the flora of beds overlying the uppermost tillite, it ended at about the same time as in those regions. It includes a far greater number of glacial and interglacial periods, as one would expect from its much longer history. The ice sheets appear to have come from lands to the south, of which most of the area is now beneath the sea, only Tasmania remaining above water. Süssmilch has suggested the name Tasmantis for this almost vanished continent.

In the other continents the glacial centers were in regions which have remained dry land since early times and are still above the sea.

One is struck by the fact, as previously mentioned, that the ice moved toward the equator, from a latitude which one supposes would be cold; whereas in India and South Africa the reverse was the case, the ice moving polewards from the tropics. Australian land ice, so far as known, ended in the Maitland region in about lat. 33°, while both the other continents had ice sheets which extended ten or fifteen degrees nearer to the equator.

On the other hand, floating ice reached much farther from the edge of the land ice in Australia than it is known to have done elsewhere, glaciated blocks being found in water-formed beds in lat. 24° in West Australia and lat. 20° 30' in Queensland.

The barrier of ice extending far out into Ross sea from the Antarctic continent is probably forming similar aqueoglacial deposits at the present time, while more sporadic dumps of glaciated stones are being dropped in lower latitudes by thawing icebergs which may reach even to lat. 40° or 35°.

Taken altogether there is more variety in the glacial deposits of Australia than elsewhere, and the wide development of laminated varve clays in New South Wales and to a less extent in Victoria and West Australia indicates the formation on a grand scale of glacial lakes. These may have equalled the glacial lakes of the North American Pleistocene in area, though their boundaries can never be very certainly determined.

The widespread uniformly stratified shales of the southern Dwyka in South Africa probably indicate similar marginal lakes of even greater dimensions before as well as after the main advance of the ice. Actual varve clays have been reported from the neighborhood of Nondweni in Natal by Du Toit,[22] and seasonally banded rocks of the kind probably occur elsewhere but have not been definitely described. In India only one or two references suggest that varves were formed.

Typical glaciated stones are fully as common in the Australian tillite as in the Pleistocene boulder clays of the northern hemisphere; but facetted stones, like those from the Salt Range in India, appear to be rare.

Effects of the Glaciation on the Flora and Fauna

Before the earliest glaciation at the base of the Kuttung series in the Middle Carboniferous the flora corresponded to the Lower Carboniferous of Europe and North America and included many lycopods, such as *Lepidodendron, Ulodendron,* and *Cyclostigma,* some equisetaceous plants, e.g., *Archæocalamites,* and a conifer, *Pitys.* Ferns are absent, probably because the plant remains are drift material consisting of stems enclosed in marine deposits.[12] One may suppose that the climate of the Lower Carboniferous was mild if not warm, since pitys, which was related to the araucarian pines, was quite a large tree, as shown by its stumps. Lepidodendron, which is the commonest plant fossil, is very frequent also in the northern coal measures.

Then came the first glaciation sweeping away the warm-climate vegetation, followed by an interglacial flora in the Kuttung series containing especially certain ferns, *Rhacopteris* and *Aneimites,* but not the *Glossoptoris* and *Gangamopteris* so characteristic of coal seams following glaciation in India and South Africa.

Above the Branxton tillite another series of coal-bearing rocks occurs at Newcastle including no less than ten seams, one reaching fifty feet in thickness; and here *Glossopteris* is the commonest plant, the flora resembling that of the Indian and South African coal fields. On dryer ground *Dadoxylon* and other trees grew, some of the stumps and trunks still standing above the coal seam.

The fauna of the Lower Carboniferous of New South Wales includes, according to Süssmilch, about 130 species of marine invertebrates, largely brachiopods but with a few other shellfish, trilobites, corals, etc. After the first glaciation many of them disappear and only a few occur in the mud of the sea bottom into which floating ice was dropping boulders. *Fenestella, Eurydesma, Aviculopecten, Keenia* and *Platyschisma,* found under these conditions, must have been hardy forms capable of living in a glacial sea.

In the Upper Marine series (Permian) an unusual number of giant invertebrates occur, which, it is suggested, heralds their approaching extinction. A labyrinthodont, *Bothriceps,* appears at the horizon of the Newcastle coal measures, but no reptiles have been reported from rocks of this age in Australia. Amphibians and a few insects are the only known air-breathing animals.

Similar relations are found in Tasmania and in West Australia, but the marine fauna of the latter state consists largely of different species, since a land barrier running north and south across the continent separated the epicontinental seas of the two regions.

Permo-Carboniferous Glaciation in New Zealand

The neighboring islands of New Zealand, which reach farther south even than Tasmania, might be expected to show the effects of any great extension of the Antarctic ice, such as the Australian tillites indicate, but until recently no evidence to that effect has been discovered. In 1920, however, James Park reported the finding of striated boulders in rocks of this age at Taieri Mouth in the South Island.[23]

Up to the present no fossils of the Glossopteris flora have been found, so that the real relation of this breccia to the tillites of the other parts of the Southern Hemisphere can hardly be considered established.[24]

Possible Glaciation in the Malay States

J. B. Scrivenor, in an account of the Gopeng Beds of Kinta (Federated Malay States), compares them with "drift composed of till and boulder clays.[25] The boulder beds carry tin ore and consist of clay through which stones are scattered, thickly or thinly. They are of all sizes, a few reaching tons in weight. No striated stones have been found and no fossils either of plants or animals are associated with the deposits, which are usually indistinctly stratified.

REFERENCES

1. *Geological Notes of a Journey in S. Austr. from Cape Jarvis to Mt. Serle,* Adelaide, 1859.
2. *Geol. Sur. Victoria,* 1866.
3. *Austr. Ass. A. Sc.* Vol. II, 1890, p. 452.

4. DAVID, *Quar. Jour. Geol. Soc.*, Vol. XLIII, 1887, pp. 190, etc.

5. *Jour. Geol.*, Vol. 20, pp. 199-214.

6. *Geol. Sur. India, Recs.*, Vol. XIX, 1886, p. 44.

7. *Quar. Jour. Geol. Soc.*, Vol. LII, 1896, pp. 289-301.

8. *Proc. Roy. Soc. Victoria, New Series*, Vol. V, pp. 45-68.

9. *Austr. Ass. A. Sc.*, Vol. V, 1893, pp. 376, etc.

10. DAVID, "Geol. of Hunter River Coal Measures," Mem. No. 4, *Geol. Sur. N.S.W.*, 1907.

11. SÜSSMILCH, *Geol. of New South Wales*, 1914, p. 83, etc.

12. SÜSSMILCH and DAVID, *Proc. Roy. Soc. N.S.W.*, Vol. LIII, 1920, pp. 246-337.

13. *Roy. Soc. N.S.W.*, Vol. XLVI, pt. 2, 1921, pp. 259-262.

14. R. M. JOHNSTON, *Proc. Roy. Soc. Tas.*, 1884, p. LXV.

15. T. B. MOORE, *Proc. Roy. Soc., Victoria*, 1894, Vol. VI, pp. 133-8.

16. "Gondwana System in Tasmanien," *Sitz. K. Boehmische Gesellschaft, etc.*, 1888, pp. 584-654.

17. *Roy. Soc. N.S.W.*, Vol. LIII, 1920, Table opposite p. 302.

18. *Rep. on Bowen R. Coal Field*, Brisbane, 1879, p. 7.

19. *Austr. Ass. A. Sc.*, Vol. VII, 1898, p. 109.

20. *Ibid.*, Vol. XIII, 1912, pp. 203-9; and *An. Rep. Geol. Sur. W. Austr.* 1900, p. 98.

21. *Proc. Lin. Soc. N.S.W.*, 1921, Vol. XLVI, Part 2, pp. 260-2.

22. *Trans. Geol. Soc. S. Africa*, Vol. XXIV, 1921, p. 206.

23. "On the Occurrence of Striated Boulders in a Palaeozoic Breccia," *Trans. New Zealand Inst.*, Vol. 52, 1920, pp. 107-8.

24. *Geol. Sur. N. Z.*, Bull. No. 23 (N.S.), 1921, pp. 29-30.

25. *Quar. Jour. Geol. Soc.*, Vol. LXVIII, 1912, pp. 140-163.

Many other references to Permo-carboniferous glaciation in Australia are to be found in successive volumes of the *Australian Ass.*, especially in reports of a commttee on glaciation. Such references occur in 1895, 1898, 1901, 1902, 1904, 1907, 1908, 1912, and 1913.

CHAPTER IX

PERMO-CARBONIFEROUS GLACIATION IN SOUTH AMERICA

Introduction

In 1888 Waagen, while discussing the palæontology of the Indian Permian, expressed surprise that one continent, South America, had escaped glaciation at that time. "The presence of a mild climate in this quarter is proven by the existence of coal measures with genuine Carboniferous plants in Brazil." He thinks that "the great revolution which occurred also in the fauna of the seas may be referred . . . to the great depression of temperature which at the end of the Palæozoic time appears to have spread itself over the whole earth, South America excepted." [1]

Up to that time no evidence of glaciation had been found, but in the same year O. A. Derby described from Capivary in southern Brazil a conglomerate with boulders of granite, gneiss and other rocks, sometimes a meter in thickness, and accounted for it as glacial. [2] He found no striated stones, however, and for many years his suggestion was overlooked except by Waagen himself who mentioned the matter in Part 2 of the Indian Survey Records in 1889.

No further reference was made to glaciation in South America until 1906 when I. C. White briefly recalled the matter [3] in a discussion of the coal deposits of southern Brazil. He classifies the coal-bearing rocks into three series, of which the lowest, the Tuberão series, includes a boulder bed, the Orleans conglomerate, followed by sandstones, shales and coal with the *Glossopteris* and *Gangamopteris* flora, evidently a similar succession to that of the Indian, South African and Australian Permo-carboniferous.

In the following year David White, who as palæobotanist had worked up the fossil plants collected by I. C. White in Brazil, gave an excellent discussion of Permo-carboniferous Climatic Changes in South America, in which he quotes I. C. White as confirming the glacial origin of the basal boulder conglomerate and draws a parallel between the Brazilian relations and those of the continents bordering on the Indian Ocean.

In 1908 I. C. White's monumental Relatorio Final appears, in which a detailed account is given of the coal beds and their related rocks in southern Brazil. In this the glacial character of the boulder conglomerate is taken for granted, but no reference is made to the final proof of glacial action in the form of striated stones,[4] so that there was still room for doubt as to the interpretation of the boulder beds.

All doubt was finally removed in 1912 by the publication of J. B. Woodworth's account of a Geological Expedition to Brazil, etc., in which typical tillite was described including many striated stones. His detailed descriptions and photographs of the tillite and of beautifully glaciated stones are entirely convincing.[5]

It is rather surprising that the excellent geologists who studied the Brazilian Permo-carboniferous beds before should not have picked up examples of these stones, but they were not glacial geologists and had other matters in view.

Woodworth was aided by two young Brazilian geologists, Drs. E. P. de Oliveira and J. Pacheco, assistants of Dr. O. A. Derby, who was then Director of the Brazilian Geological Survey. Since Woodworth's valuable report was published the discovery and mapping of extensions of the glacial deposits have been continued by Oliveira and Pacheco and the known area of tillite has been considerably increased.

In 1917 the present writer visited South America with the object of studying the Permo-carboniferous glacial deposits and the following account of the Brazilian glaciation will be from my own observations. In this work Woodworth's report was of the greatest service and I had the good fortune also, through the courtesy of the São Paulo Geological Survey, to have the guidance and assistance for part of the time of Dr. Pacheco.[6]

Glacial Beds in São Paulo, Southern Brazil

At the offices of the Geological Survey of the State of São Paulo in the city of the same name a program was worked out to visit several typical outcrops of tillite in the state under the guidance of Dr. Pacheco, who had mapped a considerable part of the glacial beds. Our first visit was to the thriving city of Campinas, about sixty miles north of São Paulo, where the deeply weathered brown or red materials due to weathering in a moist and tropical climate were not at all suggestive of ice action. In a railway cutting a mile or two from the city tillite was seen resting upon a *moutonnée* surface of gneiss. Although weathering had gone too deep to leave a polished or striated surface, the arrangement was just what one sees in northern Ontario, the hummocky gneiss underlying a sandy boulder clay enclosing subangular blocks of quartzite, granite, etc. According to Dr. Pacheco this is the only place in the state where the tillite has been found resting on a floor of ancient rock. Above the boulder bed there are shales and sandstones evidently belonging to the same series of deposits.

Our next visit was to the neighborhood of Capivary, where glacial action had been suggested many years before by Derby, though he found no striated stones. A walk of eighteen kilometers through charming tropical scenery, where plantations of rice, sugar cane, pawpaws and oranges suggested anything but glaciation, brought us to a number of tillite sections of the most characteristic kind. At one point where the wheels of ox carts loaded with cotton had cut deeply into a hillside road many striated stones were found and a bag full collected, so that we had better fortune than Derby. Colored people riding mules into the city looked at us with amused interest.

This section was as satisfactory as any I have seen, though the ten or twelve feet of boulder clay rested, apparently conformably, on sandstone and shale, sometimes cross bedded, and not appreciably disturbed by the passage of the ice.

At Villa Raffard station, four kilometers from Capivary, there are large boulders of an ancient conglomerate containing pebbles of jasper, very like some Huronian conglomerates. The source of these boulders is not known. This section was visited by Woodworth, who gives a picture of one of the boulders.

In several places in this region shales underlying the tillite have been thrown into small folds, probably by ice thrust at the time of glaciation; and sometimes a loess-like bed of fine-grained brown material which overlies it forms low vertical cliffs and is pierced by holes apparently due to decay of roots.

The tillite weathers spheroidally but when crumbled and moistened by rain it works into a slippery clay in the same way as boulder clay, though its reddish or yellowish color is not like that of our usual Pleistocene till.

Near the small station Elias Fausto the tillite contains very large granite boulders, one of which measured 9 or 10 feet in diameter. This occurrence resembled one in South Australia, and except for the weathered condition of the boulders, might be compared with the granite masses of the Iowan till sheet in North America.

Boulder of Granitoid Gneiss from Permo-Carboniferous Tillite Elias Fausto, Sao Paulo, Brazil. Dr. Pacheco Beside It.

My last excursion under the skilful guidance of Dr. Pacheco was to Limeira, fifty kilometers northwest of Campinas and a degree within the tropics, where we tramped a round of twenty-one kilometers under a hot sun and found tillite with striated stones at a number of places, at one point reaching a thickness of 75 feet. Most of it was chocolate colored, but in other respects where crumbling under the weather it closely imitated modern till, and it was hard to believe that the rock was as old as the Permian. That we were within the tropics was manifest when in one place we saw an army of leaf-cutting ants making a green band a yard wide across the red road.

It was mid-winter, July 12, but sometimes too hot in the sun; which made the thick boulder clay and beautifully polished and striated stones all the more impressive. The day reinforced the impression of tremendous climatic change which had come over me in studying the till and striated surface on Penganga River in India three years before, though there the heat was greater.

The places visited with Dr. Pacheco are all within the tropics, São Paulo itself being just to the south of the line; and tillite has been found still farther north, the known area of outcrops extending for 500 kilometers from northeast to southwest, with a width of from 50 to 100 kilometers, as estimated by the geologists of the São Paulo Survey.

Tillite in States Southwest of São Paulo

After the admirable introduction to the study of Brazilian glacial deposits provided by the kindness of Dr. Pacheco there was little difficulty in recognizing the characteristic appearance of the tillite, and on the journey by rail from São Paulo to Montevideo in Uruguay some of the localities described by Woodworth were visited, the first just beyond the southwestern boundary of the state of São Paulo between

Itarare and Sengens. In railway cuts near Sengens, Woodworth had found striated stones and large boulders of sandstone; and a walk along the railway between the two stations
proved extremely interesting. Following the crooked nar-
row-gauge railway from Sengens northeast toward Itarare
tillite is seen for eleven kilometers resting usually on sandstone, occasionally with a hummocky surface and in one case
with a suggestion of furrowing in a direction from southeast to northwest or *vice versa*. The sandstone is still soft,
and when the tillite was deposited may have been softer, so
that large blocks could easily be lifted and inclosed in the
glacial materials. In addition to these masses of local rock
there are quite large boulders of shale and of granite, and
a multitude of smaller stones, many of a harder sandstone
than the underlying rock, and a few of quartzite. The tillite
varies in thickness, sometimes reaching ten meters. Parts
of it near kilometer 241 have been more or less pushed and
crumpled, and not far to the northeast is the great fault and
escarpment mentioned by Woodworth.

The best display of tillite is about at kilometer 235, where
the smaller stones are very frequently striated, more so than
in any other till I have seen, whether Pleistocene or older.
Many of the glaciated stones show not only "soles" but well-
defined facets, as if they had been firmly held till a face was
ground flat and then adjusted at another angle, resulting in
another flat face. These facets sometimes come together
sharply. In early days similar facetted stones from the
Permo-carboniferous tillite of India attracted attention. It
would seem as if the Permo-carboniferous ice sheets held
their imbedded stones more firmly than those of the Pleistocene. Were their bases colder or was there a greater thickness of ice, giving a stronger pressure?

As may be seen from the train, tillite extends several
kilometers on the route southwest; but the next stop was

made at Ponta Grossa, midway across the state of Parana, where I. C. White had described outcrops of glacial conglomerate. On the side of the ridge on which the town is built, cuttings, made for streets and for drainage, disclose reddish, sandy glacial deposits containing subangular stones of various kinds, a few of which were found to be striated. A fairly good section is seen also on a road leading into the country. Above the tillite there is a sheet of trap weathering into a very red soil, and beneath it sandstone followed by black shale from which Devonian fossils are reported.

A visit was made also to Serinha, 70 or 80 kilometers to the southeast, where Woodworth suspected an older tillite. Typical boulder clay is passed between Palmeira and Nova Restingua and may be seen at Porto Amazonas. There is a rapid descent from Palmeira to Serinha, which is in a deep river valley at the base of sandstone cliffs. The tillite here takes the form of blue or yellow shale, readily weathering to clay containing subangular stones, chiefly sandstone, quartzite, and granite. No striated stones were found, but the bed looks like a glacial deposit. It is overlain by 200 feet of firm sandstone resembling the rock found beneath the tillite at higher levels. Beyond this fact no clue to its age was observed. The whole series of rocks, including the two tills, seems to lie nearly horizontal, doubtless with a gentle dip northwestward following the regular trend of the stratification in southern Brazil. The tillite at Serinha looks no older than that described before, and may represent merely a Carboniferous forerunner of the more important glaciation to follow.

Southeast of Ponta Grossa the railway lies too far west to give opportunities of observing the glacial deposits, passing over trap sheets, Triassic sandstones, etc.; but I. C. White's account of the boulder conglomerates associated with a low grade of coal and Permian plants in the state of Santa

Catharina shows that tillite continues to latitude 28°. His map of the Tubarão series, which includes the Orleans glacial conglomerate, extends the tillite to the southern end of Brazil, in Rio Grande do Sul, though his account does not specially mention boulder conglomerates as having been observed in that part of the country.

Suggestion of Tillite in Uruguay

In 1912 C. Guillemain described a thin sheet of tillite from the headwaters of the Fraile Muerto River in Uruguay. It was found in an abandoned shaft sunk in search of coal and had a thickness of about six feet. He found polished, scratched and grooved stones and believed the bed to be typically glacial. He gives parallel sections of the Fraile Muerto beds as compared with I. C. White's Rio Bonita beds, in which the tillite occupies the position of the Orleans conglomerate, believed by White to be glacial.[7] Associated with the tillite are banded shales, which may be interpreted as varve deposits with annual layers of coarser and finer materials.

The work of Guillemain has been criticised by Karl Walther, who is not satisfied as to the glacial origin of the conglomerate, and who quotes Woodworth as accounting for White's Orleans conglomerate as of aqueous and not glacial origin.[8] The presence of varves is suggestive of the vicinity of an ice sheet, so that the conglomerate may be aqueo-glacial if not actually a tillite.

If the Fraile Muerto beds are glacial the evidence of ice action is extended southwards to about lat. 33°.

Glaciation in Southeastern Argentina

There is a long gap between the Permo-carboniferous deposits of Brazil and northern Uruguay and the nearest outcrops of tillite discovered in Argentina, which are in the

Sierra de la Ventana not far from Bahia Blanca. Dr. J. Keidel, chief of the Geological Section of the Argentine Survey, was good enough to plan an excursion to this locality for me. A rail journey of 537 kilometers southwest from Buenos Aires brings one to the small station among the hills, after passing a vast stretch of prairie-like pampas with few or no outcrops of rock. The Sierra rises as rocky ridges with deep valleys between, one of them followed by the river Sauce Grande and others by its tributaries.

The railway crosses the river just south of the station and follows up the valley of a small stream in the Arroyo Negro. The best exposures of tillite are found in the railway cuttings along the Arroyo within seven kilometers of Sierra de la Ventana, and these will be described first.

The unweathered tillite is dark, bluish gray and entirely different in appearance from the usually red or brown and much-decayed tillite of Brazil. The rock is hard and shows some slaty cleavage, and the stones scattered through it are often a little squeezed or broken and slightly step-faulted. The weathered tillite is greenish or yellowish and crumbles somewhat readily, setting free the inclosed stones, but from the unweathered rock it is difficult to extract them unbroken. The fresh tillite is very like that from some outcrops of the Dwyka in South Africa, where the rock has undergone squeezing and distortion in mountain-building operations; and it closely resembles the Huronian tillite of Cobalt and might easily be taken for it in hand specimens.

The pebbles and boulders inclosed include several species of rocks, granites and hard sandstones being commonest. They are seldom more than half a meter in diameter and have the characteristic shapes of glaciated stones. A considerable number have well-striated surfaces and are typical products of ice action.

In some of the cuttings cross-bedded quartzite and more

or less water-formed conglomerate occur interbedded with the tillite, and in several places quartzite overlies the tillite conformably. The base of the tillite was not seen in the railway cuttings, and a search was made for it to the north, where a small stream flows toward the Sauce Grande, but in vain. On this stream the tillite has been squeezed into schist conglomerate with a marked cleavage, reminding one of the Timiscaming and Doré conglomerates of Ontario. A search still farther north showed no solid rock for several kilometers until the base of the northern range of hills was reached, where quartzite, mica schist, and slate were encountered.

Sections were examined a few kilometers up the river from the station and several fresh-looking outcrops of tillite were found at the water's edge. Ascending the slopes from such outcrops one finds weathered tillite for a few hundred yards, then a cliff of tillite, followed by a covered belt where only quartzite pebbles can be seen for a height of about 15 or 20 meters. A second cliff of tillite reaches 85 meters above the river and is followed by quartzite to the top of the ridge. The lower bed of quartzite seems to be interglacial, corresponding to the band of quartzite and water-formed conglomerate seen in the railway cuttings.

A section a kilometer or two down the Sauce Grande shows no base to the tillite, which has a thickness of 90 meters, as determined by aneroid, and is covered by quartzite including a band of tillite.

An excellent account of the glacial deposits of Sierra de la Ventana is given by Keidel in La Geologia de las Sierras de la Provincia de Buenos Ayres (1916), and the statement is made that the origin of a number of the inclosed boulders is unknown. Keidel puts stress on the resemblance of these deposits to the Dwyka, but gives no proofs of their age except that they are later than the Devonian, as shown by the inclusion of pebbles of limestone with Devonian

Striated Boulder and Tillite, Near San Juan, Western Argentina.

fossils. The hard and somewhat metamorphosed character of the rock, which seems to suggest a greater age, is to be accounted for by the action of orogenic forces.

Tillites near San Juan in Western Argentina

Following a plan suggested by Dr. Keidel an excursion was made to exposures of tillite in western Argentina somewhat south of San Juan. The nearest point to the outcrops on the railway between San Juan and Mendoza is at Paradero, kilometer 489. The railway traverses a desert country covered with sand and stones with isolated hills of rock not far to the east, and the loftier Chico de Zonda, a range of foothills of the Andes, about eight kilometers to the west, as shown on Stappenbeck's geological map of the region. Walking westward over the desert from the railway there is a gentle rise for two or three kilometers, followed by low ridges between profound ravines, apparently cut by temporary streams due to cloud-bursts in the mountains. At about five kilometers west a greenish-gray shaly or slaty rock occurs, crumbling to fine débris on the surface, and including one or two bands of dark-brown pebbles and larger stones. Most of the stones are fairly well rounded, as if rolled on a beach or in a river, and many have been broken and recemented. Frequently they have been broken again where they lie on the surface, probably by alternations of heat and cold.

A number of these stones are striated, often on more than one face. The largest seen was half a meter or somewhat less in diameter and was strongly scored. The stones are mainly basic eruptives, quartzite or limestone, the last too much attacked to show marks of glaciation. The series seems to be tilted, but the dip and the limits of the boulder bed could not be sharply determined, and in places two boulder beds occur separated by a few meters of shale.

Outcrops of loose, striated stones were followed for nearly a kilometer in a southerly direction, running parallel to the strike of the rocks in the foothills.

Somewhat to the southwest, where the narrow valley is steep walled and approaches the cliffs, a side ravine disclosed an absolutely different section, in which a boulder conglom-

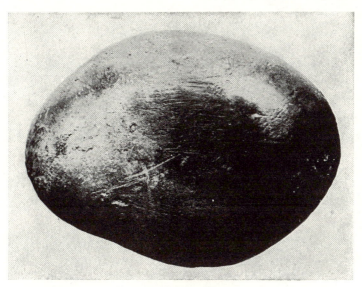

Striated Stone from Permo-Carboniferous Tillite, San Juan, Western Argentina.

erate rudely stratified in parts rises as a ridge about 30 meters high. This is of a kame-like character and includes sand, gravel, and stones of all sizes up to a meter in diameter. They are often rounded, but may be of various shapes and consist of many kinds of rocks—granite, gneiss, quartzite, vein quartz, sandstone, and limestone having been observed. Striated stones seem rare, only one poorly marked one having been found. It may be remarked, however, that in

Pleistocene kames also it is unusual to find distinctly striated stones.

The two types of deposit just described are as different as possible, though both seem to be glacial, but I was unable to determine how they are related to one another, since there has been folding, faulting, and squeezing during the formation of the mountain range, rendering the relationships complicated.

The foregoing account of the South American glacial deposits, as observed during a visit in 1917, has been given in some detail, since comparatively little information has been available in previous writings. Since then Dr. Keidel has published two important reports, covering the Argentine tillites and their associated rocks as well as comparing them with the Dwyka and other regions glaciated in Permo-carboniferous times. In 1921 he gave an account of the geology of the Pre-cordillera, as the Andean foothills have been named, in which a striated surface of the Devonian rock beneath is mentioned. His pictures of this and of striated stones from the tillite are typical.[9]

His later work I have not seen, but from a somewhat detailed review in the Geologisches Centralblatt he has evidently carried his investigations much farther than in 1921.[10]

The review states that in northwest Argentina he has found Permian tillites associated with marine deposits, indicating that the ice, coming from the east, reached the Pacific at sea level; but the geology along the Andean foothills has been greatly complicated by later mountain foldings and thrust planes. Mention is made also of the recent discovery of Permian-Gondwana beds in Paraguay. These include layers with marine fossils.

Keidel comes to the conclusion that there were two glacial centers, a northern one including the Brazilian and northwest Argentinian tillites, and a southern one including those of Patagonia and the Falkland islands; and that these were separated by a Permian sea.

The plant beds overlying the tillites are described and the flora compared with that of India; and Keidel also emphasizes the relations of the glacial beds and the trends of the South American mountains with the corresponding features of South Africa. He thinks of the two continents as blending at that time into a vast continent of Gondwana-land.

As marine beds are found with the tillites both in Brazil and in southwest Africa, it seems to me much more probable that the two continents were separated by the sea at that time, as they are now.

Glacial Conglomerates in Bolivia

Recently Dr. Mather has reported Permo-carboniferous glacial beds from southern Bolivia, probably a continuation of the Pre-cordilleran glaciation of western Argentina. Mather and Heald visited Bolivia in the interest of an oil company, so that their study of the tillite was but brief and incidental.

Glaciated pebbles occur in the Bermejo sandstone, a deposit of fluvio-glacial origin; and there is also a true tillite at one point.

The beds with striated and facetted stones are considered to have been formed mainly by glacial streams coming from ice-covered highlands, though the beds themselves seem to have been formed on low ground, since they are followed by marine sediments.[11] In describing the Mandiyuti conglomerate Mather states that "quartzite, quartz, chert, and granitic igneous rocks are the most common kinds inclosed in the matrix. Most of them are well rounded and polished by stream or current action, but in every locality there are many angular pebbles with facetted faces. Some of these are unquestionably shaped by glacial action. Most of the pebbles are small, but stones three or four inches in diameter are not rare, and in certain places boulders as great as three feet in length have been observed. The formation is evidently a fluvio-glacial accumulation, composed of débris spread broadcast over a lowland area by streams issuing from melting ice. . . . Doubtless it is genetically related to the tillite in the Bermejo beds, observed along the trail from Cochabamba to Santa Cruz, and to the glacial formations in the eastern Andes in northern Argentina, recently described by Keidel as being overlain by strata carrying the Gondwana flora." [11]

The conglomerate is from 1,500 to about 3,200 feet thick, and represents long continued glaciation.

Cochabamba and Santa Cruz are about in lat. 18°, showing that the glaciation extended even farther toward the equator in Bolivia on the west of South America than in Brazil on the east. If the Bolivian area was connected with the areas in Argentina and Brazil it is evident that the South American ice sheet of the Permo-carboniferous time was on a very large scale.

Permo-carboniferous Glaciation in the Falkland Islands

At about the time I. C. White and David White were bringing out their important report on the Coal Regions of Southern Brazil, in which a basal conglomerate was proved to be almost certainly glacial, J. Halle published a brief account of the geology of the Falkland Islands in which Permo-carboniferous tillite was reported.[12] The age was fixed by the finding of plants belonging to the *Glossopteris* flora.

In 1912 a more detailed account of the geology of these islands was published by the same author.[13] The tillite is an unstratified mudstone with pebbles and boulders of granite and sandstone up to half a meter in dimensions. A few are striated and some are facetted. No striated floor was found on the Devonian rocks beneath.

Above the tillite there are laminated shales (varves) with annual layers from a few millimeters to a centimeter in thickness, making up about two meters. Above this there are slates and sandstones with a little impure lignite and plant remains including *Phillotheca, Glossopteris, Gangamopteris* and *Dadoxylon* with annual rings. A few insect remains but no vertebrates have been found in the upper beds.

The tillite and other beds, named the Lafonian series, are found in about lat. 52°; much farther south than the tillites on the mainland, which reach only 38° at Sierra de la Ventana in Argentina.

More recently H. A. Baker has discovered a striated surface beneath the tillite and has proved that the ice sheet came from the south.[15]

The Falkland Islands, or Malvina Islands, as the South Americans prefer to call them, although more than 300 miles east of the nearest point on the mainland, are upon the continental shelf and are in all respects except the intervention of a shallow sea an integral part of the continent. The tillites found upon these islands may therefore be included with those of South America in summing up our knowledge of the glaciation of that continent.

Area and General Features of Glaciation in South America

From the foregoing account it will be seen that tillites of the Permo-carboniferous occur at many points between lat. 18° and lat. 52°, a stretch of 34 degrees or about 2,380 miles from north to south; and are found in the state of São Paulo in Brazil as well as in the middle of Bolivia and north of San Juan in western Argentina, a stretch from east to west of more than 1,400 miles.

It is improbable that these dimensions really indicate the outer limits of glaciation, since in Brazil the deposits now outcrop near the edge of a tableland which has been eaten back for a long distance since Triassic times. The glaciated area of Brazil may even have had its center to the east of the present continent in the region now covered by shallow seas. The Archæan boulders in the tillite appear to have come from the east and the source of the jasper conglomerate boulders is unknown unless they came from much farther to the north.

Du Toit, following Wegener's hypothesis of continental drift, even suggests that South Africa and South America were joined in one continent at the end of the Palæozoic.[14] The late Dr. Branner in a private communication (July 1918) expressed his belief that there are evidences of glacial deposits well within the state of Minas Geræs, which would extend the area to a latitude on the east equivalent to that shown by Mather in Bolivia toward the west.

It may be expected that further exploration in a continent relatively so little known will much increase the number of occurrences of tillite and will show that glaciation took place still farther to the north.

Even the known limits indicate glaciation at more widely separated points in South America than in Australia or South Africa and leave India far behind.

The known outcrops of tillite in South America are, however, more thinly scattered than in South Africa, except in the State of São Paulo, and striated surfaces beneath the tillite are unknown except in northwestern Argentina and the Falkland Islands, suggesting a less severe glaciation than in South Africa and Australia.

Interglacial beds have been mentioned by Woodworth and

the present writer, but are seldom found and are unimportant as compared with those of Australia.

Varve clays or slates seem to be rare on the mainland of South America, though the shales mentioned by Mather as composed commonly of beds an eighth to a quarter of an inch in thickness in the Cuestas de Oquita, Bolivia, may be of this nature. At this point there are six to eight hundred feet consisting "chiefly of thin beds of fine grained sandstone and sandy shales which display a variety of tints." [11] As mentioned before, they occur in the Falkland Islands.

Up to the present there is little to indicate whether the whole of the area outlined above was covered by a single great ice sheet 2,000,000 square miles in extent or whether there were two separate sheets, a northern and a southern, as suggested by Keidel. The ice reached sea level in Brazil, Paraguay, Argentina, and probably Bolivia, and was certainly of the continental type.

Relations of the Flora and Fauna to Glaciation

Typical Carboniferous plants and animals have not been described from rocks beneath the Permo-carboniferous tillites of South America; where, except for local sandstones and shales without fossils, the underlying beds are Archæan or Devonian. The substructure is marine Devonian in parts of Brazil, the Sierra de la Ventana, northwestern Argentina and Bolivia; so that a long period of dry land conditions must be assumed before the ice sheets began their work.

Oliveira reports from Passinho south of Imbituvo "a small *Lingula*, scales of fishes, and remains of undetermined brachiopods and lamellibranchs." On the Rio Negro, which he visited with Woodworth in 1908, he found a black combustible shale with *Lingula, Discina,* fishes and remains of sponges. In a ravine near T. Soares he found *Lingula, Discina, Chonetes* in abundance and other brachiopods, scales of fishes and wings of insects.[5] The insects must have been brought down by rivers or else must have been drifted in by the wind and drowned.

The plants associated with impure coals a little above the tillite in Southern Brazil have been elaborately reported on by the Whites, and no detailed account of them need be given here.[4] In the Tuberão series shales and sandstones with coal, just above similar beds with the Orleans conglomerate, contain the *Glossopteris* and *Gangamopteris* flora. It is essentially identical with the corresponding flora in India, Australia, and South Africa. In the higher beds northern Permo-carboniferous types are introduced and "appear to indicate the progress of a great climatic change."

The absence of Lycopods in the lower flora of Brazil indicates cold. They "come in and increase as one ascends until predominant in coals 157 meters above the granite floor."

The extermination of the northern flora was due to the refrigeration. From the Passa Dois beds, near the top of the Tuberão series, *Dadoxylon* also has been obtained. The plants of the *Gangamopteris* flora belong to families well known in the so-called cosmopolitan flora, but are simpler in figure with a tendency to thickness and rugosity of leaf.

In the Iraty black shale in Parana about 100 meters above the coal measures with *Glossopteris,* I. C. White found *Mesosaurus brasiliensis,* related to the South African species, and at the same horizon in São Paulo, Derby found *Stereosternum tumidum.* Three hundred meters higher there are red beds containing reptilian remains related to *Euskelesaurus* of the Stormberg beds in South Africa.[3] It will be seen that both the plants and the animals following the cold period in the continents of the southern hemisphere are related to one another, a fact which suggests land connections of some kind, perhaps by way of the Antarctic continent, where the *Glossopteris* flora has been found, but no animals nor tillite.

REFERENCES

1. *Rec. Geol. Sur. Ind.*, Vol. XXI, part 3, p. 129.
2. "Spuren einer Carbonen Eiszeit in Südamerika," *Neues Jahrb. f. Min.*, Vol. 2, 1888, pp. 172-6.
3. *Science, N.S.*, Vol. XXIV, 1906, pp. 377-9.
4. *Relatorio Final*, 1908, published in Rio Janeiro with alternate pages of Portugese and English. David White's account of the plants is given in full.
5. "Geol. Exped. to Brazil and Chile, 1908-9," *Mus. Bull., Harvard*, Vol. LVI, No. 1, pp. 46-82 and plates 20-27.
6. "Permo-carboniferous Glacial Deposits of S. Am.," *Jour. Geol.*, Vol. XXVI, pp. 310-324.
7. "Beiträge zur Geologie Uruguays," *N. Jahrb. f. Min*, Beilageband 33, p. 234, etc.
8. *Zeitsch. des Deutschen Wis. Vereins, Argentiniens*, pp. 386-7 (no date).
9. "Observaciones Geologicas en la Precordillera," *Geol. Sur., Argentina, Monografias*, Buenos Aires, 1921, Tome XV, No. 2.
10. *Geol. Centralblatt*, Band 30, 1924, pp. 400-405. Review 1061, by Jaworski, of Keidel's publication Sobre la Distribución de los Depósitos Glaciares del Permico, etc., *Bol. Acad. Nac. de Cienc. de Cordoba*, Vol. 25, pp. 239-368.
11. *Bull. Geol. Soc. Am.*, Vol. 33, 1922, pp. 563-5 and 734-762.
12. "Note on the geology of the Falkland Islands," *Geol. Mag.*, Vol. V, pp. 264-5.
13. "On the Structure and History of the Falkland Islands," *Upsala Univ. Geol. Inst., Bull.*, Vol. II, pp. 115-229.
14. *Trans. Geol. Soc. S. Af.*, Vol. XXIV, 1921, pp. 219-221.
15. "Review of Final Rep. on Geol. Investigations in the Falkland Isds.," *Geogr. Jour.*, Vol. LXV, 1925, p. 73.

The first vague suggestion of ice work in South America was made by CASTELMAN in—*Expedition dans les Parties centrales de l'Am. du Sud.*, Vol. I, pp. 276-8, Paris, 1850.

CHAPTER X

Boulder Conglomerates in Nova Scotia

In North America, as in Europe, comparatively few
instances of glacial deposits of Permian or Carboniferous
age have been described and no large area of tillite has been
found. The first reference to the subject seems to have been
made by Sir Wm. Dawson in 1872, when he called attention
to the New Glasgow boulder conglomerate, in Nova Scotia,
which he says "seems to be a gigantic esker, on the outside of which large travelled boulders were deposited,
probably by drift ice, while in the swamps within, the coal
flora flourished and fine mud and coaly matter were
accumulated." [1]

As shown on the sheets of the geological map of Nova
Scotia, the conglomerate extends for seventy miles with a
width in one place of eight and a half miles. Its greatest
thickness is given as 1,600 feet.

Two visits were made to examine the New Glasgow conglomerate and it was found to have much the character of
a kame or a series of kames. Its materials are coarse,
including more or less rounded boulders of several kinds of
rock of all dimensions up to two or three feet—rarely as
much as six feet.

A few striated stones have been found, but probably all
due to internal motions among the pebbles. Except at one
point, near Tatamagouche, there is little fine material and

New Glasgow Conglomerate, Nova Scotia.

no typical till; yet the whole effect of this great mass of coarse, partially assorted material suggests glacial outwash and morainic conditions.

In the absence of any known range of mountains to provide torrential deposits on a large scale it seems natural to account for the New Glasgow or Tatamagouche boulder conglomerate as due to the combined work of ice and water.

The conglomerate has been mapped by some as Upper Carboniferous and by others as Permian. So far as reported it is non-fossiliferous, but plentiful coal plants of the cosmopolitan flora occur with coal seams at a lower horizon, and a small coal seam is found just above it.

There are coarse boulder conglomerates on a much smaller scale at the base of the Lower Carboniferous of the same region, sometimes containing blocks several feet in diameter, but no one has suggested that they are glacial. If they should prove to be glacial the Pictou coal measures would be interglacial like the Greta coal near Newcastle, New South Wales.

A much more careful examination of these conglomerates will be necessary before one can be sure of their glacial character. Sir Wm. Dawson described the New Glasgow conglomerate in his Acadian geology without any suggestion that it was glacial, and in his brief statement, quoted above, from the *Canadian Naturalist,* he gives no decisive proofs of its resulting from the action of ice.

A Possible Permian Moraine in Prince Edward Island

Probably the next suggestion of glaciation of this age in North America was made in 1886, when F. Bain briefly described a supposed Permian moraine at the north end of Prince Edward Island. "At Blackman's Island a conspicuous ridge or mound ten to twenty feet in height, fifty yards broad, and five hundred yards in length . . . has much the appearance of a Quaternary moraine. . . . This ridge of ancient conglomerate is composed of rounded masses of red sandstone sometimes 2 feet in diameter, gravel, sand, and clay . . . without any stratification."

No striated stones are mentioned and no later geologist has referred to the supposed moraine, which occurs in land deposits, mainly red sandstones, mapped as Permian by the Canadian Geological Survey.

The two supposed glacial deposits of eastern Canada deserve attention mainly because they are in line with the well authenticated tillite at Squantum near Boston, a few hundred miles to the southwest, which will be taken up next.

The Squantum Tillite

The boulder-bearing beds near Boston have attracted the attention of geologists for many years, as was to be expected from their nearness to a great city and a great university, but final proof of their glacial origin was not really accomplished till R. W. Sayles undertook a careful and detailed study of the various outcrops.[2] Probably no other set of ancient glacial deposits has been worked over so thoroughly as those of Squantum. All the characteristic features of till are shown in the various outcrops and are elaborately demonstrated by Sayles; so that a complete account of them is unnecessary here. The character of the matrix, the nature and size of the enclosed pebbles, boulderets, and boulders, and the presence of striated stones are described and figured, and the glacial nature of the deposits is convincingly proved. Having visited some of the outcrops with Prof. Woodward and Dr. Sayles, I can testify that the rock has exactly the appearance of the Permo-Carboniferous tillites of other continents.

The Squantum rock is a typical ancient boulder clay and not kame like, so that it does not greatly resemble the boulder conglomerates of Nova Scotia, mentioned heretofore; and it is more certainly glacial than the New Glasgow conglomerate. The tillite is distributed over an area of about 100 square miles.

Associated with the tillite near Boston there are extensive beds of seasonally banded slate, varves, like the Pleistocene glacial lake deposits of Sweden and the eastern United States and Canada. These have been excellently described by Sayles,[3] and reinforce the proofs of ice work afforded by the tillite itself. Sayles deserves much credit for calling attention to the probable aqueo-glacial origin of such banded slates and shales in different ages of the past.

Alaskan Permo-Carboniferous Tillites

The only other region in North America from which probable glacial deposits at the end of the Carboniferous or in the Permian have been mentioned is on the Alaska-Yukon boundary and in Alaska itself.

In 1914 D. D. Cairnes reported from the boundary, just north of Tatunduk River, an area of a square mile of supposed tillite consisting of a "firm, somewhat dense, finely textured, reddish, argillaceous matrix, in which are embedded angular to subangular pebbles and boulders ranging in size from microscopic to 3 or 4 feet in diameter. The matrix appears to have approximately the composition of boulder clay." [4]

The boulders are chiefly limestone or dolomite but some sandstones, conglomerates and shales occur also. No striated pebbles were found but some of them have facetted surfaces much resembling "soled" pebbles. Cairnes believes that the boulder bed, which is 700 or 800 feet thick, was formed about in Carboniferous time and "may correspond to the Permo-Carboniferous tillites of South Africa, India and Australia."

In 1918 Edwin Kirk reported a similar tillite from Alaska, which he correlates with Cairne's boundary conglomerate, and which overlies beds high in the Carboniferous and underlies the middle Triassic. These outcrops are on Pybus Bay, Admiralty Island, and on the Screen Islands.[5]

Other Possible Permian or Carboniferous Tillites

A boulder conglomerate described and figured by Whitman Cross from West County, Colorado, south of Dolores River, may be glacial, though this is not suggested in the description. The photograph is quite glacial looking, and the deposit is given as probably Permian.[6] Those western Red Beds were of continental origin and deserve scrutiny as possible glacial deposits.

Glacial boulder conglomerates have been reported from the Caney shales, Talihina, Oklahoma; but these may belong to the end of the Mississippian or the beginning of the Pennsylvanian. As described by J. A. Taff there are boulders, cobbles and rock fragments of limestone, flint, chert and quartzite, of all sizes even up to fifty feet in length, some of them being striated.[7]

Woodworth, who examined the boulder deposit in 1912, come to the conclusion that the striæ are due to interstitial motion, but that the boulders may have been transported by floating ice. Some of them are fifty miles from their source.[8]

In 1923 Samuel Weidman found both striated boulders and striated surfaces beneath in an old U-shaped valley. His illustrations show scoured and grooved surfaces that suggest glaciation and support his conclusion that a mountain glacier did the work.[9]

Twenhofel has accounted for certain granite boulders, found far south of the Pleistocene glaciation, as coming from a Pennsylvanian shale, into which they were dropped by floating ice when the shale was still mud;[10] and Shaler and Davis have suggested that in many places in eastern America boulder conglomerates (the Pottsville, etc.) may have been glacial;[11] but this view does not seem to be approved by geologists in general.

Carl O. Dunbar, in an article on Permian Insects from Kansas, doubts all of these conclusions and accepts only the Squantum tillite as a real glacial deposit. He remarks that "over against such evidence of glaciation the testimony of the fossil animals and plants of the time is quite over-whelming. There is probably no other period of the past for which the biologic evidence for warm equable climate is so unmistakable as the Pennsylvanian." He quotes David White's statement that "the climate of the principal coal-forming intervals of the Pennsylvanian was mild, probably near tropical or subtropical, genially humid and equable."[12]

From the account given on previous pages it is pretty certain that the usual idea of tropical conditions during the Carboniferous should not be urged too strenuously. It has recently been shown by Winnifred Goldring that trunks of *Cordaites recentium* from Oklahoma and other regions in North America have annual rings;[13] so that there was a change of seasons and not the continuous steamy heat supposed to go with coal swamps.

The finding of undoubted glacial beds in New South Wales, as reported by Süssmilch and David, in formations corresponding to the Pennsylvanian proves that this time was cold in the southern hemisphere;[14] and the Indian tillites indicate the same conditions north of the equator at the end of the Carboniferous.

The argument from the "cosmopolitanism" of the coal measures flora is not worth much when one recalls that these rank growths occurred only in parts of North America and Europe and are scarcely known from any of the other continents. Instead of being cosmopolitan they have a peculiarly local distribution, centering about what is now the north Atlantic. The coal of the other regions was formed by plants adapted to the comparatively cool climate which seems to have prevailed in the rest of the world.

It is possible that the relatively slight cooling of North America and Europe, as compared with the extreme cold of the other continents, may be accounted for as due to more oceanic conditions and warm currents directed northwards. The southern continents and India may have had land connections (Gondwana land) which cut off these warming influences.

The idea formerly held that the latter regions were glaciated because of their elevation as tablelands has been disproved by the finding of glacial deposits at sea level in all of the continents affected.

The Permo-carboniferous ice sheets seem to have begun on peneplained surfaces, probably not very far above sea level, and to have spread widely on the low ground, showing little deflection by the moderate elevations of the land forms encountered; and in many cases they actually reached the sea. In Australia the free ice margin probably expanded far over the sea before breaking up into icebergs.

In regard to their origin on low ground and their reaching and encroaching upon the shallow seas these Permo-carboniferous ice sheets of the Southern Hemisphere closely paralleled those of the Pleistocene in the Northern Hemisphere. The area of glaciation in the last ice age almost exactly covered the parts of the world which escaped glaciation at the end of the Palæozoic.

There were local mountain glaciers in Australia and the tropics and piedmont ice sheets in Patagonia during the Pleistocene ice age; just as there were small glaciated areas in Europe and North America in the Permo-carboniferous.

Correlation of the Late Palæozoic Glacial Deposits

In the accounts given of the glaciation of the different regions suggestions have been made of parallel features of the deposits themselves and of the plants and animals associated with them; and as far as India, South Africa, Australia and South America are concerned the parallelism is impressive.   The tillites themselves have frequently been used as defining a horizon for correlation purposes, and there has been a tendency to relegate all the tillites of those regions to the same horizon.   It has been suggested, for example, that the Katanga tillites and others in Central Africa were all of Dwyka age, instead of Triassic; and that the Blaini tillite beds of the Himalayas were of Talchir age, though later work has proved them to be much older.

It is evident that such correlations, especially when extended from continent to continent, imply a theory of the causation of ice ages.   There is an implication that the causes which produced the late Palæozoic glaciation were world-wide in their operation, or at least affected the whole of the Southern Hemisphere and the Indian Peninsula.

The lively and long continued controversy as to the age of the Talchir tillites in India, in which palæobotanists made them Mesozoic and palæontologists Palæozoic, was one of the most interesting features of the early investigation of these ancient boulder clays, and illustrates the difficulty of determining the geological period of a land formation.

The chronology of historic geology is essentially founded on the gradual evolution of the marine faunas of the world. This is partly because life probably began in the shallow seas, which have always been more prolific than the land; but mainly because the ocean has been a unit since the earliest known times, so that new species could migrate to remote shores; and also because the marine record has been continuous, though some of the connecting links may be unknown to us.

Tillites are typical land formations and from their mode of origin, which excludes living beings, can have no indigenous fossils. The few remains of plants and animals contained in a true tillite must have been derived from older formations.

Interglacial beds, when the cold relented sufficiently to allow of life, may help in determining the age of a glacial series; and fossils in beds succeeding the glaciation without a break are very helpful; but the land forms preserved in these deposits are often of little use for age determination. The inhabitants of the land have much more individuality than those of the sea; and the flora and fauna of a given period may differ entirely on different continents.

Under the circumstances the Permo-carboniferous age of the tillites remained a matter of dispute until marine deposits including striated stones dropped from melting ice were found to contain also fossils of marine animals. This happened first in the New South Wales coal region. In other cases tillite was found between marine beds containing fossils. Similar relationships have been found in all the southern continents, as well as in India, so that there is no longer any doubt as to the correlation, so far as those regions are concerned.

In addition the *Glossopteris* or *Gangamopteris* flora, consisting mainly of hardy ferns, has been found in sandstones or shales interbedded with the tillites or immediately following them in the regions just referred to. These plants did not reach North America or Western Europe until a much later time, and then in a much modified form. It is remarkable that such a flora, relatively poor in species and coming so soon after the departure of the ice or sometimes even in an interglacial period, should have formed beds of coal in all the chief glaciated regions.

The relations of the glacial and post-glacial floras have been excellently discussed by David White in several papers and in I. C. White's Relatorio Final, prepared for the Brazilian government, and it will be unnecessary to go into details in this work.[15, 16, 17]

The latest and most complete and suggestive correlation of events in the Permo-carboniferous glaciation of the different regions has been given by Sir Edgeworth David. He shows in eight parallel columns the succession of glacial, interglacial, and immediately post-glacial deposits of the time in New South Wales, Victoria, Tasmania, West Australia, South Africa, India (Olive series of Salt Range), South America and North America (Squantum).[18] The sections are of very unequal magnitudes, that of the Hunter River District N.S.W., including no less than 17,430 feet while that of the Salt Range in India covers only 75 feet.

In all of the sections more or less interglacial stratified material is shown, but in the Hunter River Section there are thick interglacial coal seams implying long and relatively mild periods intervening between ice ages; while in South Africa, India, and at Squantum in North America, the enclosed stratified beds are comparatively thin and are usually devoid of fossils.

There is no evidence that the long and mild interglacial times of Australia had an equivalent in the other regions, and it is probable, as suggested by Süssmilch and David, that glaciation began there long before the other continents were affected.

In considering the causes of the Permo-carboniferous glaciation it is necessary to keep in mind the fact that Southern Australia was ice covered hundreds of thousands, or perhaps millions of years, before the other regions showed signs of glaciation. Was this because it was more directly exposed to the invasion of Antarctic ice?

In the case of the other great period of glaciation, the Pleistocene, there is no evidence of ice sheets attacking one region far in advance of the others, though what occurred in Antarctica and Greenland before the beginning of the Pleistocene ice age is unknown. It may be that in those regions, which are practically out of reach of exploration, there was Pliocene glaciation and then a milder period before the Pleistocene ice sheets began to spread out over Northern Europe, North America and Patagonia.

Effects on Life of the Late Palæozoic Ice Age

References have been made to the effects of the ice sheets upon the plants and animals of the continents which were heavily glaciated; but it is desirable to show briefly the more widespread results of this most tremendous of the world's ice ages. The Palæozoic, the time of ancient life, ends with the Permian, when most of the formerly dominant types of living beings disappear or lose their importance, leaving the way open for new types to take their place. This is true of sea and land and air. The great ice sheets reached the sea on several continents, and in Australia, at least, set free fleets of icebergs to chill the waters of the tropics. It was too serious an ordeal for many creatures adjusted to warm waters, and we find that trilobites vanish, corals and brachiopods diminish greatly, and few of the many primitive sharks of the Palæozoic seas survive. The antiquated ganoid fish with bony scales or plates almost disappear.

In their place come the more modern and adaptable mollusks, ammonites, bony fish and great marine reptiles.

On the land the giant spore-bearing plants, horsetails, ground pines and tree ferns, lose their supremacy and give place to conifers and cycads, the forerunners of Mesozoic forests. Among the cryptogamic trees there were many strange insects, including forms like dragon flies with a two foot spread of wings. The climax in size of the lower forms of plant and animal life, the spore plants and the insects, passed with the long winter of the Permo-carboniferous ice age, leaving the way clear for the flowering plants and flying vertebrates, such as the pterodactyls and the birds with teeth, of the Mesozoic. On the land the small reptiles which survived the cold rapidly multiplied and expanded into the dinosaurs which ruled the renovated continents after the ice sheets disappeared.

W. D. Matthew, one of the best known students of vertebrate palæontology, says that "the period was a most important and critical one in the evolution of land life, for it witnessed the first great expansion of land vertebrates and the origin, probably, of mammals, birds, and the principal orders of reptiles, including dinosaurs." [19]

The Permo-carboniferous glaciers piled up much thicker sheets of boulder clay and moraine, and also included more important interglacial deposits than the Pleistocene. It is not surprising that the life of the world suffered great losses during these changes of climate, clearing the way for great advances when the glacial hardships were over.

But for this rude interruption possibly gigantic insects with large brains might have led the world intellectually in later times instead of vertebrates. Knowing how efficient and ruthless are the instincts of our tiny modern insects the thought is somewhat appalling. Their line of advance, so far as size is concerned, seems to have been cut off by the advent of frigid conditions; and the next group of animals to develop mightily was the reptiles.

If the small dip in temperature which occurred in the Eocene had not halted the progress of the cold-blooded saurians and destroyed the largest and most advanced of them, who knows that the dinosaurs or pterodactyls might not have led the way up to the present instead of warm-blooded, well-clothed, mammals and birds?

Ice ages have occasioned far-reaching and salutary changes in the world's population as judged from the human point of view.

REFERENCES

1. *Canadian Naturalist and Geologist,* Vol. VI, p. 416.
2. "The Squantum Tillite," *Bull. Mus. Comp. Zoöl., Harvard,* Vol. LVI, No. 2, 1914.
3. *Proceed. Nat. Acad. Sc.,* 1916.
4. *Geol. Sur. Can.,* Mem. 67.
5. *Geol. Soc. Am.,* Vol. 29, pp. 149-151.
6. "Colorado Red Beds," *Jour. Geol.,* Vol. XV, pp. 662-5.
7. *Geol. Soc. Am.,* Vol. 20, pp. 701-2, "Ice-borne Boulders."
8. *Geol. Soc. Am.,* Vol. 23, pp. 457-462, "Caney Shales, Oklahoma."
9. "Pennsylvanian-Permian Glaciation in the Arbuckle and Wichita Mountains of Oklahoma," *Jour. Geol.,* Vol. XXXI, 1923, pp. 466-489.
10. *Am. Jour. Sc.,* Vol. 43, 1917, pp. 363-380.
11. SHALER and DAVIS, *Glaciers,* 1881, pp. 97-99.
12. *Am. Jour. Sc.,* Vol. VII, 1924, pp. 190-196; and DAVID WHITE—"The Origin of Coal," *Bull. Bur. Mines,* No. 38, 1913.
13. *Botanical Gazette,* Vol. LXXII, 1921, pp. 326-330.
14. "Carboniferous and Permo-carboniferous Rocks of N.S.W.," *Proc. Roy. Soc. N.S.W.,* Vol. LIII, pp. 246-338.
15. "Carboniferous Glaciation in Southern and Eastern Hemispheres," *Am. Geol.,* Vol. 13, 1889, pp. 299-332.
16. "Permo-carb. Climatic Changes in S. Am.," *Jour. Geol.,* Vol. 15, pp. 615-633.
17. F. H. KNOWLTON, "Evidences of Palæobotany as to Geological Climate," *Science, N.S.,* Vol. 31, 1910, p. 760.
18. *Roy. Soc. N.S.W.,* Vol. LIII, 1920, p. 302.
19. "Recent Progress and Trends in Vertebrate Palaeontology," *Bull. Geol., Soc., Am.,* Vol. 34, 1923, p. 404.

CHAPTER XI

Devonian Glaciation

GLACIATION in the Devonian is found in comparatively few regions and on a small scale as compared with that of the later Palæozoic just described. It has been suggested in Silesia, in England and Scotland, in Alaska, in the northeastern United States, in eastern Canada, and in South Africa; the only large area being in the last-mentioned country. In the others the known deposits may be accounted for by mountain glaciers rather than lowland ice sheets.

One of the first suggestions of Devonian glaciation is probably that of F. Rogner in 1870.[1] He found in Upper Silesia rocks of this age including clay slate which enclosed granite boulders up to several hundred weight, and alternate sheets of graywacke and slate an inch or less in thickness, apparently varves.

In 1875 Sir Charles Lyell referred to "supposed signs of ice action in the Old Red Sandstone, or Devonian Period," and mentions that in 1848 the Rev. J. G. Cumming, in his History of the Isle of Man, compared the conglomerate of the Old Red to a consolidated boulder clay. He also quotes Ramsay as mentioning "stones and blocks distinctly scratched, and with longitudinal and cross striations, like the markings produced by glacial action." There were pebbles of all sizes up to two feet or more, but the rocks had undergone much faulting and some slickensides occurred, so that Lyell concludes that there should be more evidence before one would be convinced of their glacial origin.

James Geikie states that "the Scottish conglomerates often present the appearance of morainic débris, being frequently unstratified, while the stones show that peculiar subangular blunted aspect which is characteristic of glacial work. In these deposits my brother, Sir A. Geikie, has detected many striated stones." In the unstratified parts "the rock masses are confusedly huddled together in a tough arenaceous matrix, and the accumulation then closely resembles a boulder clay. It is typically developed in the Lammermuir hills, but similar deposits are met with in Ayrshire and other regions in the west of Scotland." [3]

Certain coarse conglomerates in the valley of the river Samme in Belgium have much the look of glacial deposits. They belong to the Poudingue d'Alvaux in Middle Devonian. The rock is brown and contains pebbles of all sizes up to one and a half feet in longest diameter. They are subangular and rounded and often have glacial looking shapes. Several showed faint striations, but a careful comparison with typical striated stones makes it probable that the markings are due to adjustments in the mass itself. The boulder conglomerate is worthy of a more careful examination than could be given on a rather hurried excursion of the Geological Congress of 1922.

There are no recent references to Devonian glaciation in the Old World, so far as my reading goes, and it would be useful if some geologist familiar with the Permo-Carboniferous tillites would reëxamine the localities mentioned above and determine whether the deposits are really glacial or not.

The Alaskan peninsula seems to have had a cool climate during most of its history as shown by Blackwelder from the gray and dark colors of its sedimentary rocks. Organic matter has remained undecayed even in ancient sediments because of cool and moist conditions.[4]   A probable glaciation at the end of the Carboniferous has already been mentioned; and Blackwelder refers to a bed of tillite containing striated subangular boulders in connection with certain shales and cherts "perhaps of Devonian age."   Edwin Kirk, also, mentions probable glacial deposits in the Middle Devonian on the west coast of Prince of Wales Island and in Freshwater Bay and Chicagoff Island, some 250 miles to the north.   At the latter place there are rounded boulders up to two feet in diameter.[5]

From its position in the far north glaciation might be expected in any general cooling of the earth, but much of Alaska is little known geologically and one can only suggest probable glaciation in the Middle Devonian.

In the eastern United States there are few references to ice action in the Devonian.   J. M. Clarke has described certain rod-like markings and groovings on sandstones of the Upper Devonian of New York as probably caused by ground ice;[6] and E. S. C. Smith gives an account of the Rangeley conglomerate in Maine, which is associated with banded beds, the whole being perhaps of aqueo-glacial origin.[7]

The evidence from eastern Canada seems more definite. J. M. Clarke in 1914 described a coarse conglomerate at Scaumenac on the south side of the Gaspé peninsula in Quebec, of which he says "some of the pebbles show unqualified evidence of glacial scratching and the entire mass is regarded as an outwash from glacial moraine."[8]

In 1918 I visited the locality and can confirm the conclusion reached by Clarke.   The stones are of all sizes up to a foot and a half in diameter and are generally well rounded as if by river action.   A few are distinctly striated.   It is probable that the Shickshock mountains, a few miles to the

Boulder Conglomerate, North of Hampton, New Brunswick.

north, were then much higher than now and sent down glaciers which formed the deposit.

The age of the bed is fixed by the well-known fish beds of the Upper Devonian which overlie it. The latter beds were laid down in fresh water and the whole series of rocks was of continental origin.

The most striking boulder conglomerate supposed to be of Devonian age in eastern America was described by G. F. Matthew in 1920 from the neighborhood of St. John, New Brunswick.[9]

Here a boulder conglomerate extends for twenty miles from northeast to southwest and has a width of eight miles, with a thickness of 240 feet in one place. Boulders of various kinds with diameters up to three feet are imbedded in an unstratified matrix of clay, sand and pebbles. A few striated stones are reported, but from personal observation I am inclined to think them due to rearrangements in settling and not to ice action. The conglomerate resembles a moraine or kame deposit on a large scale and can hardly be accounted for otherwise than by the work of a glacier.

The Devonian age of the conglomerate is not absolutely certain, since no fossils occur directly associated with it and the palæobotanists call the nearest "fern ledges" Lower Carboniferous; though the field geologists who have worked in the region put it in the Upper Devonian.

The most important development of Devonian glacial beds is found in the southern part of Cape Colony, where Rogers and others have worked out an extensive series of conglomerates containing many striated stones of a typically glacial character.

The Table Mountain Series, mainly of sandstones, contains at Cedarbergen 300 feet of mudstone, the lower hundred feet unstratified and containing scattered stones of several kinds of rock, some of them well striated. There is no

sharp boundary between the boulder bearing lower portion
and the shale above, which is finely laminated, so that apparently the tillite passes up into a set of varve shales.   No
striated floor is found beneath the tillite, which Rogers
accounts for as formed by floating ice.

The tillite has been followed for twenty-three miles and
is probably much more extensive than this; and a layer of
sandstone or quartzite enclosing well striated stones has
been found on Table Mountain itself more than a hundred
miles farther south, so that the whole area then covered by
glacial materials was large.[10, 11, 12]

The age of the Cedarbergen and Table Mountain glaciation is not very certain, since no fossils have been found in
the Table Mountain Series or the beds below, which were
all, so far as known, land formations.   The Bokkeveld
Series, which overlies it conformably, contains Devonian
fossils, so that the glacial deposits are probably Lower
Devonian; though they may be Upper Silurian.   Schuchert,
no doubt from a consideration of the fossils in the overlying Bokkefeld, which he calls Lower Devonian, makes the
glacial deposits Silurian.[13]

Silurian Glaciation

The most important examples of Silurian ice action,
unless the instance cited above is of this age, occur in
Alaska, where Kirk has found glacial deposits on Heceta
Island, on the south shores of Kosiusko Island, and on the
northern parts of the Kuiu islands 125 miles to the north.

The most striking development of tillite is on the north
shore of Heceta Island, where a boulder conglomerate with
a thickness of 1,000 or 2,000 feet lies between two limestones containing plenty of marine fossils.   The length of
the exposure is 2,000 or 3,000 feet.   There are boulders
of greenstone, graywacke and limestone, as well as various

types of igneous rocks. Facetted boulders are common and fine-grained dense greenstone boulders are often striated. Limestones, however, are not. Kirk believes the deposit to be true till and not marine beds containing ice-transported blocks.[14]

Professor Blackwelder, also, has found outcrops of tillite, probably of Silurian age, in several places in Alaska. In a personal communication he mentions three instances. On the Yukon River, at the mouth of the Fourth of July Creek, there is a bed about 80 feet thick, associated with black slates and dolomites. "The tillite is typical, unstratified boulder clay, folded but not metamorphosed. The associated beds can hardly be younger than Middle Palæozoic and it is thought likely that they are Silurian."

Perhaps the best determined tillite on the Yukon occurs eight miles below Woodchopper store. "The material is an unstratified, olive-gray shale with incipient slaty cleavage. The pebbles and boulders of all sizes are scattered in the usual aimless fashion. About four or five specimens with very well-preserved striæ were discovered, although in most cases the matrix adheres to the surface of the pebble. The bed is about 300 feet thick and excellently exposed. It is in contact on one side with cherty dolomite containing poorly preserved fossils, apparently of Silurian, but possibly of Devonian age."

At the Lower Canyon of Beaver River on the south edge of the Yukon flats there are extensive exposures of slate enclosing boulders of many kinds and of all sizes up to ten feet or more in length. The boulder slates occur in a thick series of intensely folded sedimentary rocks, the age of which is not closely determinable, though they are probably Palæozoic.

If all of the Alaskan occurrences are Silurian, one may infer that a large area was ice covered at this time, implying a sheet of the continental type rather than the piedmont glaciers, which are so characteristically developed in Alaska at the present time.

Nine hundred miles south of Alaska on Sinclair River and Shuswap Creek in southeastern British Columbia, F. P. Shepard has found a Silurian boulder conglomerate with stones as large as four feet. The thickness of the conglomerate is about two hundred feet and it extends about ten miles between the two streams. Some of the boulders are subangular and facetted and a few are probably striated. Some of the pebbles have grooved and concave faces. The boulder bed is enclosed in marine deposits on both sides, the whole series of beds having been tilted till nearly vertical.[15]

In an inaugural address as President of the British Association in 1880, A. C. Ramsay refers to glaciation in Wigtonshire and Ayrshire, in Lower Silurian rocks; but these beds would now be placed in the Ordovician. He mentions. also boulder beds with indistinctly ice-scratched stones on the Lammermuir hills, south of Dunbar, Scotland, as resting unconformably on Lower Silurian strata, which would put them at the base of what is now called Silurian.[16]

Medlicott and Blandford mention that "old slates, supposed to be Silurian, contain boulders in great numbers," in the Himalayas near Rangi, southeast of Kashmir.[17]

The Silurian tillites are never very widespread, though one of them, that of Heceta Island in Alaska, is very thick. It is probable that all of them except those of Alaska can be accounted for by local mountain glaciers. The Alaskan tillite, 1,000 or 2,000 feet thick, is enclosed between marine rocks; but great glaciers reach the sea in that latitude at the present time and may have done so then.

The presence of great beds of salt in the Silurian of Ontario and the neighboring states is generally held to imply a hot and dry climate; but Schuchert states that toward the close of the Silurian there was a great change in the life of the seas indicating a considerably reduced temperature so that "even local glaciation may have been present." [18]

It may be that in reality the Silurian and Devonian refrigerations are parts of the same period of cooling which marked the transition from the one age to the other.

Ordovician Glaciation in Europe

In the older literature the only references I have found to Ordovician glaciation are by A. C. Ramsay and James Geikie. The former mentions boulder deposits in graptolitic rocks of the Lower Silurian (Ordovician) at Corswell Point, Wigtonshire, and at Carrick in Ayrshire. There are erratic blocks of all sizes up to nine feet in length, consisting of gneiss, granite, etc., which must have been transported from a distance, probably by icebergs or other floating ice.[18] No mention is made of striated stones; and it might be desirable for some geologist familiar with glacial matters to look over the deposits again. James Geikie mentions these Scottish deposits also.

Glaciation has recently been described by Olaf Holtedahl from Ordovician rocks of far-northern Europe. In Finmarken he finds two sheets of tillite, an upper one ten meters thick at Bossekop and Mortensness, and a lower one two or three meters thick at Bigganjargga. The latter rests conformably upon sandstone which is nicely striated. The upper one lies between shaly sediments, from which it is separated by no distinct boundary. There are striated stones in the tillite, one which is figured being very characteristic.

As no fossils are found in the series of sandstones containing the tillites their age is uncertain, but Holtedahl thinks they are possibly of late or middle Ordovician age.[19]

The tillites referred to the Ordovician by Holtedahl were

described years ago by Reusch and Strahan as belonging to the Gaisa system, usually supposed to be of early Cambrian or late Pre-cambrian age.[20, 21]

The Limestone Conglomerate

By far the most extensive and important glacial deposits of Ordovician age have long been known to Canadian geologists as "limestone conglomerates," because of the great

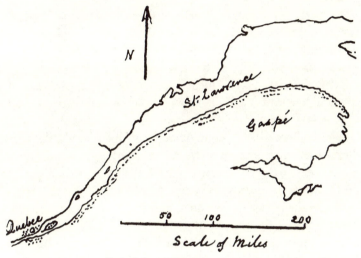

Ordovician Tillite in Quebec.

number of limestone boulders found in them and also because the matrix is usually an impure limestone. Though they have often been referred to and described since the earliest days of the Canadian Geological Survey, they have usually been mentioned without any attempt at an explanation of their origin. The latest account of them in the Reports of the Survey, by L. W. Bailey and W. McInnes, shows that the boulders have their only known sources at great distances away. These writers account for them as

probably "originally laid down along an extended shore line
defined approximately by their present distribution, for the
formation of which the materials were carried down from
the north by streams and rivers." [22]

They note with surprise that there are no granites or
gneisses among the boulders, though the Laurentian highlands rise only a little way to the north. The Laurentian
must have been covered with Cambrian limestone at the
time.

In 1918, while engaged in Pleistocene work in Gaspé, I
was struck with the glacial look of the boulder conglomerate,
but found no striated stones and made no mention of the
matter until 1921.[23] In the same year, by an interesting co-
incidence, R. W. Sayles mentioned the glacial appearance of
the conglomerate near the city of Quebec and also the fact
that it is associated with varves.[24]

The boulders in the conglomerate include, beside the predominant limestones, basalts, amygdaloids and sandstones.
The blocks are of the usual glacial shapes and of all sizes
up to twelve or even twenty-five tons. The limestone boulders contain Upper Cambrian fossils and their nearest known
sources are at the Straits of Belleisle or at Lake Mistisinni,
250 or more miles away. The origin of the basic igneous
boulders is unknown. The nearest basic eruptives of a
similar kind are on the other side of the Gaspé Peninsula
70 or 80 miles to the south. It is evident that the transport
of boulders of the dimensions just given can hardly be
accounted for by river action unless mountain torrents were
at work, and lofty mountains were not in existence in the
region at that time, so far as known.

The beds of boulder conglomerate are of various thicknesses up to 150 feet, and often two boulder beds are separated by shale or sandstone. Occasionally there are several
bouldery layers with intervening stratified materials; and, as

Boulders of Ordovician Tillite, Rouisseau Loutre, Quebec.

shown by Sayles, varve shales occur with them in some places near Quebec and Levis.

These limestone conglomerates have been followed from Quebec city to the end of Gaspé peninsula, a distance of 360 miles; though there are a number of relatively short interruptions. The boulder beds are almost always steeply tilted, since they were involved in the thrusts which formed the Appalachian mountains. The greatest known width is ten miles; but before the folding and faulting in mountain building they must have extended much more widely. They are scattered over several thousand square miles at present but probably covered a far greater area at the time of their formation.

No striated stones nor striated floors have been found, as might be expected in rocks which have undergone mountain-building stresses; but in every other respect the boulder beds have a glacial character and many outcrops have the appearance of typical tillite.

The great area covered, the large size of the blocks, the great distance from which some of them must have come, and the fact that the Laurentian mountains had long before been worn down to a peneplain, while the Appalachians had not yet begun to rise, limit one to glaciation as the only possible cause of these boulder conglomerates. The absence of mountains, just mentioned, and the fact that marine fossils are found in rocks above and below indicate that the work was done by an ice sheet of the continental type on low ground.

An interesting corroboration of this conclusion was found in an old Canadian Geological Survey Report while looking up the extensive literature on limestone conglomerates. A. P. Low reports of the Ordovician rocks of the Quebec region that the fossils contained in them are of diminutive size, implying very cold water, and that a block of granite eight tons in weight, embedded in the later Trenton limestone, must have been transported by floating ice.

Seven hundred miles southwest of Quebec, in the northwest corner of Vermont, Arthur Keith has described the Swanton conglomerate of Middle Ordovician age as probably glacial.[25] In many respects this closely resembles the Quebec limestone conglomerate. It contains even larger boulders and is accompanied by uniformly banded slate (Georgia slate). "The Swanton-Georgia sequence is a close duplicate of the Pleistocene till-clay sequence. The principal difference is that the Pleistocene beds contain more of the far-travelled boulders than the Swanton, a difference attributable, perhaps, to a more local origin of the Ordovician glaciers."

As both tillites were squeezed up with other sediments in the formation of the Appalachian ranges it is natural to think of them as belonging to the same period of refrigeration, but Keith considers the Swanton rock to be younger than that of Quebec.

In a private communication R. W. Sayles informs me that regularly banded Lower Ordovician slates occur at Rockmart, Ga., Johnson City, Tenn., and Melrose, N. Y. The banding is undoubtedly seasonal and is very like the varve slates of Squantum, which are associated with Carboniferous tillite.

Although no direct evidence of ice action is known from these localities, there must have been cold winters, suggesting a climatic relation to the northern tillites mentioned above.

If all of the indications of ice action and seasonal changes in the Ordovician belong to the same period it would seem that a considerable depression of temperature affected northern Europe and eastern North America at the time, the American glaciation having been much more severe than the

European and in a more southerly latitude, as in the Pleistocene.

REFERENCES

1. *Geologie von Oberschlesien,* Breslau, 1870, p. 18.
2. *Principles of Geology,* Vol. I, pp. 229-30.
3. *The Great Ice Age,* 3rd Ed., 1894, p. 818.
4. "Climatic History of Alaska from a new View Point," *Trans. Ill. Acad. Sc.,* Vol. 10.
5. *Geol. Soc. Am.,* Vol. 29, 1918, p. 151.
6. New York State Museum, *13th Ann. Rep. of Director,* 1916, p. 206.
7. *Am. Jour. Sc.,* Vol. V, 1923, pp. 147-154.
8. *Geol. Soc. Am.,* 1914, Abstract of paper; See also *13 Ann. Rep.,* mentioned above, p. 209.
9. *Roy. Soc. Can.,* Vol. XIV, Sec. IV., pp. 4, etc.
10. ROGERS refers to the Devonian glaciation in *Trans. Philos. Soc. S. Af.,* Vol. XI, 1902, pp. 236-242; in the same, Vol. XVI, Part 1, 1905; and in *Geology of Cape Colony,* 1905, pp. 94-121.
11. SCHWARZ mentions it in "Three Palaeozoic Ice Ages," *Jour. Geol.,* Vol. 14, 1906.
12. HANS CLOOS, "Die Vorkarbonische Glazialbildungen des Kaplandes," *Geol. Rundschau,* Leipzig, 1915-16, pp. 337, etc.
13. "The Climates of Geologic Time," *Carnegie Inst. Pub.,* No. 192, pp. 277-290.
14. *Geol. Soc. Am.,* Vol. 29, 1918, pp. 149, etc.
15. *Jour. Geol.,* Vol. XXX, 1922, pp. 77-81.
16. *Nature,* Vol. XXII, 1880, pp. 388-9.
17. *Geology of India,* Vol. I, p. XXXVI, and Vol. II, pp. 632 and 664.
18. *Nature,* Vol. XXII, 1880, pp. 388-9.
19. *Am. Jour. Sc.,* 4th Ser., 1919, pp. 86-106.
20. REUSCH, *Skurungs-Marken og moraena grus i Finmarken,* Christiania, 1891; and *Geol. of Northern Norway,* 1891, pp. 30, etc.
21. AUBREY STRAHAN, Glacial Phenomena of Palaeozoic Age in the Varanger Fiord, *Quar. Jour. Geol. Soc.,* Vol. LIII, 1897, pp. 137-146, with plates showing a striated surface of sandstone under the tillite.
22. *Geol. Sur. Can.,* Vol. V, 1890-91, M.
23. "Pres. Address," *Trans. Roy. Soc., Can.,* 1921, p. XLII.
24. SAYLES, *Abstract, Geol. Soc. Am.,* 1922.
25. *Am. Jour. Sc.,* Vol. V, 1923, pp. 118, etc.

CHAPTER XII

Early Cambrian or Late Pre-cambrian Tillites in Europe

OMITTING the Varanger Fiord tillites as probably Ordovician, there is still a Norwegian till-like deposit which may be Cambrian in age. This was described by Holtedahl from the "Eocambrian" Sparagmite of southern Norway in 1922.[1] The set of sandstones called Sparagmite underlies the Lower Cambrian Holmia shale without any stratigraphical break, so that its age is fixed. "The conglomerate . . . has all of the general characteristics of a tillite, and it is difficult to think of any other method of transportation for material of this kind than a glacial one. As yet, however, no undoubted glacial striæ have been observed, but since the sparagmites have suffered much from the Caledonian deformation, the finding of striæ is very difficult. In fact, when the boulders are freed from their matrix their surface is generally seen to be slickensided and tectonically striated. This conglomerate is exactly like the brown tillites of Finmarken" (described on a former page).

The sandstones associated with the conglomerate are really arkose, since they contain a large amount of fresh feldspar. Holtedahl assumes, because of this feldspar, that the materials were due to weathering of a granitic rock in a very dry climate. He is apparently not aware that unweathered feldspar grains can be formed equally well in a cold, moist climate, as may be seen at the present time in

Labrador and the granitic area of the Shickshock mountains in Quebec, where coarse angular grains of fresh feldspar and quartz accumulate on the lower ground.

At one time the Torridonian of the Highlands of Scotland was thought to include boulder conglomerates of glacial origin. Sir A. Geikie says of them, "Some of the conglomerates are so coarse as to deserve the name of boulder beds. Sometimes, indeed, where the component blocks are large and angular, as at Gairloch, they remind the observer of the stones in a moraine or boulder clay. Some of the sandstones are in large measure composed of pink feldspar derived from such rocks as the pegmatites of the surrounding gneiss." [2] Geikie compares them with the Sparagmite of Norway, and considers them the uppermost beds of the Precambrian, since rocks with Cambrian fossils overlie them unconformably.

No striated stones are recorded from the Torridonian, and later geologists, including Peach and Horne who have done such admirable work in Highland geology, consider them of desert origin. The reasons for this view are the unweathered feldspars, mentioned above, and the finding of dreikanters in some of the beds.

I spent a few days in studying the Torridonian in 1922 and found no very conclusive evidence of desert conditions in its formation. The arkose is readily accounted for by a cold and moist climate, and dreikanters may be formed under the same conditions. They are now being formed on the Magdalen islands in the Gulf of St. Lawrence. All that is required to shape them is drifting sand which may be found in dry weather on any seashore.

The whole effect of the Torridonian seems to me to indicate a cool and moist climate with shallow lakes into which coarse materials could be swept. The sandstones and shales of the Torridonian are well stratified and show no duny structures in the regions seen by me. I had no opportunity to study the Gairlock section, but suggest that some geologist familiar with ancient tillites should look the conglomerate over to see if it does not contain striated stones.

Far to the north of Europe in Spitzbergen what may be a tillite of this age was described a good many years ago by Gregory. It was found in the Hekla Hook series, which Gregory thinks probably equivalent to the Gaisa beds of Norway. No striated stones were seen, but blocks of granite and gneiss with dimensions up to seven feet by five were found enclosed in an imperfectly foliated matrix.[3]

Some later authorities believe this to be a crush conglomerate, but the finding of glacial deposits in rocks of a similar age and character elsewhere makes it worth while to reconsider the question.

Early Cambrian or Late Pre-cambrian Boulder Conglomerates in North America

The Keweenawan of the Lake Superior region has often been compared with the Torridonian and has very similar characters. It is interesting, therefore, to note that it also includes boulder conglomerates which have been interpreted as glacial. Robert Bell reported boulders reaching three feet eight inches in dimensions and having grooves like glacial striæ, in a conglomerate with sandy matrix belonging to the Keweenawan of Pointe aux Mines, near the southwest end of Lake Superior.[4]

Lane and Seaman also describe a Lower Keweenawan conglomerate from the south shore of the lake as containing "a wide variety of pebbles and large boulders, in structure at times suggestive of till." [5]

Recently a till-like boulder conglomerate has been described by Kieth from the Milton dolomite (Upper

Cambrian) in northwestern Vermont. It is followed by laminated slates like varves. No striated stones have been found in the conglomerate. The suggestion is made that there was high land in the region supporting glaciers, and the fact that the beds are of Upper Cambrian age appears to put them out of line with the other tillites referred to.[6]

A boulder conglomerate very like a tillite occurs in the Sudbury district at a somewhat lower horizon than the Keweenawan; in beds considered to be of Animikie age. The Trout Lake conglomerate is at the base of a series of sediments found resting on the sheet of nickel-bearing eruptive for which the region is famous. In most places this boulder bed has been greatly metamorphosed by the eruptive rock which spread out beneath it in the molten state; but at the west end of the nickel basin, where the eruptive sheet thins out, it is better preserved.

Near the Sultana Mine and Cameron Creek the rock consists of a dark gray, fine grained matrix enclosing pebbles and boulders of quartzite, greenstone and granite, boulders of the last rock being often three feet or more in diameter, the largest found having diameters of nine feet by twelve. The bed varies in thickness from a few feet up to 450, and is continuous round the basin, so that the known outcrops have a length of 34 miles and a width of ten and a half. It varies greatly in the number and size of the enclosed boulders.

Although the rock is much like a tillite no striated stones have been found in it and its glacial origin has not been certainly proved.[7]

The latest report of tillite from the Lower Cambrian or perhaps Pre-cambrian, has reached me in a letter from Eliot Blackwelder, who writes, "I am sending you a small specimen of tillite from the shore of Great Salt Lake, Utah.

Last summer I studied this locality in detail and finally succeeded in finding a few striated pebbles and boulders. The rock has all the other usual characteristics of tillite.

"The age of this deposit is not yet accurately determined. However, it cannot be younger than Lower Cambrian, and

Photographed by Eliot Blackwelder.

Lower Cambrian or Older Tillite, Little Mountain, West of Ogden, Utah.

it may well be Late Pre-cambrian. On the other hand, it contains abundant fragments of Early Pre-cambrian quartzite."

Photographs of the tillite and a hand specimen sent with the letter are characteristically glacial looking.

Tillites have been reported from a number of points on islands in Great Salt Lake and on its shores, and also in the Wasatch mountains. In some cases these are considered to be older than Late Pre-cambrian by their discoverers, and will be referred to later.

Early Cambrian Glaciation in Asia

In 1907 Bailey Willis and Blackwelder published an account of important glaciation, probably in the Early Cambrian, from the Yangtzi Canyon in China,[8] and since then two other parties of geologists have visited and described these deposits. The tillite is 120 feet thick and lies between gritty sandstone beds, above which is a thick marine limestone of the Middle Cambrian. The Nantou tillite has a green, somewhat sandy, matrix enclosing fragments of all sizes up to two and a half feet. The pebbles are of great variety and many of them are smoothed and striated. Specimens brought back are admirable examples of glaciated stones.

No striated floor has been found, but the tillite has the usual appearance of till formed by a glacier and not by floating ice. The glacier must have about reached sea level; and as the latitude is 31°, this implies serious refrigeration.

A later examination by Iddings confirmed the account of the previous party and proved that the area of tillite is considerable, extending for miles near Nantou and being found on both sides of the Yangtzi River. It occurs also at Lungling, where it is about 300 feet thick, but the distance between the two points is not mentioned.[9]

An ancient boulder conglomerate in the foothills of the Himalayas has attracted a good deal of attention from the Indian geologists and is referred to in several reports of the Survey. At first it was considered to be of the same age as the Talchir tillite and was used as a horizon marker in working out the stratigraphy. It was observed, however, that no fossils of any kind occurred in the series of rocks to which it belonged, unlike the other tillites, which were followed by the *Gangamopteris* flora.

In 1908 Sir Thomas Holland described the Blaini conglomerate as found near Simla and came to the conclusion that it was very much older than the Permo-carboniferous. He puts the Purana series, in which it occurs, beneath the

Photographed by Sir. Thos. Holland.

Striated Stone from Early Cambrian or Late Pre-Cambrian Tillite, Simla, India.

Vindhyan, and considers it in whole or in part Pre-cambrian. The Blaini tillite is then in all probability of late Pre-cambrian age. Holland compares it to the glacial conglomerates of the Yangtzi in China, and also to the older tillites of Australia and South Africa.

A picture of a striated stone from the Blaini at Simla is very typical and there can be no doubt as to the glacial nature of the rock from which it came.[10]

Early Cambrian or Late Pre-cambrian Tillites in South Africa

The examples of early Cambrian or late Pre-cambrian glaciation thus far considered are all from the Northern Hemisphere, and none of the areas appears to be very extensive. If they alone were in question one might account for all the phenomena by mountain glaciers and it would not be necessary to assume very important refrigeration at this time in the world's history. South of the equator tillites are much more widely spread and imply, as in the case of the Permo-carboniferous, a marked lowering of temperature amounting to a glacial period, when continental ice sheets spread over large areas in South Africa and Australia.

The first recognition of a probably Pre-Cambrian tillite in South Africa seems to have been by Rogers in 1905. He describes it as "a very well developed glacial till, often with an intensely hard ferruginous and siliceous matrix, so that the boulders cannot be broken out from it. The release of the boulders is brought about by the slow disintegration of the matrix due to alternate heating and cooling and weathering. In a few cases I was able to break boulders out of the rock, and thus prove that their beautifully striated surfaces, as seen on weathered out specimens, belong to them when in the rock."

He found the tillite in a number of places in the Griquatown series in Hay.[11]

The tillite was traced for 115 miles from south to north with a width of about 30 miles in the earliest exploration; and in 1909 the area over which its outcrops were known to occur was estimated at 8,000 square miles. No striated surface has been found beneath it, but the tillite is so much like the Dwyka that there can be no doubt as to its glacial origin.[12]

Specimens of the tillite and examples of the striated pebbles were sent me by Rogers some years ago, and I can testify to their glacial character. The views of Rogers are concurred in by Schwarz [13] and Hans Cloos,[14] who have briefly described the tillite.

The early glaciation of South Africa reached to the north of lat. 29°, i.e., approached much nearer the equator than that of the Pleistocene, and there is no evidence of great elevation of the land, though the absence of marine fossils in the Griquatown Series makes not only the age of the beds but their altitude uncertain.

As to the age of the tillite there is much doubt. It may be either Cambrian or Pre-cambrian, the only certainty being that it is much older than the Devonian, in which marine fossils have been found.

Early Cambrian (or Late Pre-cambrian?) Tillites in Australia

The first reference to Lower Cambrian glaciation in Australia was by W. Howchin in 1901, when he called attention to boulder conglomerates of that age near Adelaide in South Australia.[15] Since then his explorations, with assistance from David and other geologists, have extended the area of glaciation over thousands of square miles in that state, and finds of similar tillite have been made in Tasmania and in New South Wales.

A very good and complete account of the Lower Cambrian tillite was given by Howchin seven years later under the title of "Glacial Beds of Cambrian Age in South Australia," [16] which will be drawn upon for the following description. "The beds which give evidence of glacial origin may be described as consisting mainly of a ground mass of unstratified, indurated mudstone, more or less gritty, and carrying angular, subangular and rounded boulders up to eleven feet in diameter, which are distributed confusedly through the mass. It is in every respect a characteristic till."

The glacial series has a thickness in some sections of 1,500 feet, usually including a considerable amount of stratified material, quartzite, finely laminated slate or limestone, in one or more layers. For instance, the Appila Gorge section shows a total thickness of 1,526 feet, with an upper till of 120 feet, separated by 656 feet of stratified beds from 750 feet of lower till. Some of the stratified rocks include scattered erratics. Much of the interglacial slate appears from his description to be of the nature of varves, and the overlying Tapley Hill slates, which are described as laminated and banded, are probably of the same nature.

The enclosed boulders are of several kinds, largely granites, gneisses and schists derived from the Pre-cambrian, and two erratics of graphic granite seem to have been derived from the southern Yorke Peninsula, 122 miles from the spot where they were found. This indicates a transport from southwest to northeast. No striated surface is found beneath the tillite, so that there is no other evidence as to the direction from which the ice came.

Howchin discusses the question of interglacial periods, as suggested by the stratified beds between tillites, and does not find any widespread evidence of important climatic variations.

The lack of an underlying striated surface is mentioned as opposing the hypothesis of land ice as a cause for the tillite and, like most Australian geologists, he believes that floating ice did the work.

In 1914, under the leadership of Professor Howchin, a party of geologists who were members of the British Association had an excellent opportunity to study the tillite as exposed in the Sturt River section seven miles south of Adelaide, and during that excursion I was strongly impressed

Photographed by John Greenlees.

Early Cambrian or Late Pre-Cambrian Tillite, Sturt River, South Australia.

with the typical appearance of the boulder clay. It seemed to me to be entirely the work of land ice. In most places there was no evidence of stratification, and the varying distribution of the boulders and the considerable size of some of them gave the impression of a real till and not of a deposit with which water had anything to do.

Icebergs, in dropping their burden as they thawed, would necessarily give rise to some classification of the materials as they sank with unequal speed through the water.

It is a well demonstrated fact that the Pleistocene beds of North America were formed by land ice, yet there are thousands of square miles of boulder clay with no striated surface beneath. An ice sheet near its edge, where it is clogged with englacial materials in its lower parts, is depositing and not scouring, just as a river in slack water aggrades its channel instead of cutting it down. There is no evidence in Pleistocene glaciation that icebergs can form important deposits of typical unstratified boulder clay.

A striated rock surface beneath till is good evidence of erosion by land ice, but the lack of a polished and striated surface is no proof that a sheet of boulder clay was not deposited by land ice. In fact the thicker the mass of till the more likely it is that it was deposited beneath the almost stagnant edge of the ice.

What has just been noted applies, in my opinion, to much of the Permo-carboniferous tillite with no striated surface beneath it in South Africa as well as Australia.

The tillites just mentioned form very thick beds, probably because laid down near the edge of the ice sheet, and include beautifully striated stones, as proved by reproductions in Howchin's excellent account. He estimates that the outcrops extend for 460 miles from north to south with a width of 250 miles across the main line of strike. They reach lat. 29° toward the north.

It is evident that the glaciation was on a very broad scale, since the center from which the ice radiated seems to have been somewhere to the south of the present continent, so that the whole area may have been more than double that which has been described.

Although Australian geologists place this tillite in the Cambrian it should be remembered that only the upper limit

After W. Howchin.

Striated Boulder from Base of Cambrian, Petersburg,
South Australia.

of its possible age has been settled. There are thousands of feet of barren sediments between the till and the first fossiliferous beds, the *Archæocyathina* marbles.

Howchin points out that the tillite is separated from the "basal complex" by a less thickness of rock.

The basal complex looks to a Canadian as old as the Timiskaming Series; and it may be that the division between the basal Cambrian and the uppermost Pre-cambrian is no better marked in South Australia than in British Columbia where the dividing line between the Cambrian and the Beltian is very uncertain.

Schuchert, from the evidence of a mild climate afforded by the Lower Cambrian fossils, thinks the glaciation took place in the late Pre-cambrian.[17]

The evidence of glaciation just above or just below the boundary of the Cambrian and Pre-cambrian has been given at sufficient length to show that an important cooling affected the climates of the world at that turning point. Both hemispheres experienced a fall of temperature, but the southern hemisphere seems to have been more seriously chilled than the northern. In this respect conditions are parallel to those at the end of the Palæozoic, though the cooling seems to have been much more severe in the latter glaciation.

Although the evidence as to the extent of glaciation is naturally very imperfect in so ancient a time as the beginning of the Cambrian, it seems sufficient to prove a greater depression of temperature than that of the Pleistocene. In the ice age through which the world has just passed the nearest point to the equator reached by a continental ice sheet was about 38°, while land ice, apparently at low levels, reached latitudes both north and south of the equator of 29° or 31° at the beginning of the Palæozoic.

I know of no glacial deposits of the Pleistocene, excluding the work of mountain glaciers, which reach 1,526 feet in thickness, including 870 feet of till, so that in this respect, also, the ancient glaciation surpasses the Pleistocene.

So little is known of Pre-cambrian life that speculations as to its abundance or importance before the time of glaciation have no value; but the rapid blossoming out of all types of animals except vertebrates in the Cambrian is one of the most impressive features of palæontology. Did the cold waters of the ice age destroy some oppressive oligarchy of degenerate tyrants among the marine invertebrates and free and revivify by hardship other species so that they could multiply and expand in the warming seas?

As to plants the only ones reported from the Pre-cambrian are algæ or bacteria, and these seem to have survived all vicissitudes to the present day. The land, so far as known, was devoid of all life, an absolute desert, so that the work of ice sheets could not add to the desolation.

REFERENCES

1. *Am. Jour. Sc.*, Vol. IV, pp. 165-173.
2. *Textbook of Geology*, 3rd Ed., 1893, p. 705.
3. *Quar. Jour. Geol. Soc.*, Vol. LIV, p. 216.
4. *Geology of Canada*, 1876-7, p. 214.
5. *Jour. Geol.*, Vol. XV, p. 688.
6. *Am. Jour. Sc.*, Vol. V, 1923, pp. 113 and 132-3.
7. *Bur. Mines, Ont.*, Vol. XXIII, Parts 1 and 2, p. 234; also "The Nickel Industry," *Can. Dept. of Mines*, 1913, p. 9.
8. Carnegie Inst., *Research in China*, Vol. 1, Part 1, p. 267.
9. Quoted by SCHUCHERT, *Climates of Geologic Time*, p. 294.
10. *Geol. Sur. Ind.*
11. *Trans. Geol. Soc. S. Af.*, Vol. IX, 1906, pp. 8 and 9.
12. *Geology of Cape Colony*, pp. 96-7; also *S. Af. Ass. A. Sc.*, 1906, pp. 261-5.
13. "Three Palaeozoic Ice Ages in S. Af.," *Jour. Geol.*, Vol. 14.
14. *Geologische Rundschau*, 1915-16, pp. 346-7.
15. *Trans. Roy. Soc. S. Austr.*, Vol. XXV, p. 10.
16. *Quar. Jour. Geol. Soc.*, Vol. LXIV, 1908, pp. 234-259.
17. *Carnegie Inst.*, Pub. No. 192, p. 276.

There are numerous references to the Australian tillites in successive volumes of the *Austr. Ass. A. Sc.* from 1902 onwards, but they do not add much to Professor Howchin's account mentioned above.

CHAPTER XIII

GLACIATION IN HURONIAN AND EARLIER TIMES

Huronian Tillites

Of late years evidence of still more ancient glaciation has been growing and now one can speak with certainty of a Huronian ice age of much importance.

Boulder conglomerates, or "slate conglomerates," as they were called in the beginning, have long been known from the Pre-cambrian of Ontario and Quebec, particularly from the typical Huronian region, and roused the interest of the older geologists in Canada. Various explanations of them have been suggested, such as a torrential origin or desert weathering or even explosive volcanic action; but the present writer had been impressed with their glacial appearance for several years before final proofs were obtained that they were the work of ice. This evidence turned up first in the Cobalt silver mining region of Ontario, which may be described as typical.[1]

The Cobalt conglomerate is now placed in the Gowganda Formation of the Cobalt Series, the latter being the equivalent of the upper member of the original Huronian.

The opening of the silver mines gave excellent exposures of the conglomerate and within a year or two a number of characteristically striated stones were obtained, most of which are now displayed in the Royal Ontario Museum in Toronto.

As the finding of evidences of glaciation so low in the geological time scale seemed almost incredible in 1907, care was taken to compare thin sections of the matrix and the stones contained, in it with examples of Dwyka tillite which had been collected in South Africa a year or two before. The two rocks are strikingly alike in both hand specimens and thin sections and might easily be interchanged. Every

Lower Huronian Tillite, North Shore of Lake Huron.   (Striated Surface of Pleistocene Glaciation.)

feature proving the glacial origin of the Dwyka tillite is repeated in the rock from Cobalt.

The rocks of the underlying Laurentian and Keewatin, such as granite, gneiss, basic eruptives, and banded iron formation, occur in the tillite as fragments of all sizes up to several feet in diameter, the largest measured being eight feet by five as exposed on the surface; and their shapes are subangular, but often smoothed and "soled" in appearance. In parts of the Huronian region near Lake Huron blocks of granite occur more than twenty miles from the nearest known source.

The enclosed stones may be crowded together or widely scattered, a boulder in a square yard or more. Usually there is no visible stratification, but near Cobalt there are inter-

Striated Stone from Cobalt (or Gowganda) Tillite, Cobalt, Ont.

bedded bands of water-formed conglomerate and also of laminated slate—interglacial deposits.

In earlier years the objection was made that the Cobalt boulder conglomerate rested in places on weathered products of the rock beneath, and was nowhere underlain by a striated surface. This is not a serious objection, however, since many later boulder clays rest on weathered materials. In

1918, Mr. Burrows, of the Bureau of Mines of Ontario, described a smoothed surface of Keewatin beneath the tillite at Matachewan;[2] and in 1922 Dr. H. C. Cooke discovered a glacially gouged, polished and striated surface near Opasatica Lake in the Province of Quebec. The striæ run N. 60° E., and there is reason to believe that the ice came from the southeast. The bed of tillite is fifteen feet thick and contains granite boulders up to fifteen feet by eight in dimensions, at a point twelve miles from the nearest outcrop of such granite from which it could have been derived.[3]

Since a few geologists have recently expressed doubts as to the glacial character of the Cobalt tillite, it may be well to state that the tillite and striated stones have been examined by many of the most experienced glacial geologists of Europe and America and all have agreed that they are the work of ice.

At the meeting of the Geological Congress in Stockholm (1910) Professor Molengraaff (Delft) stated that "die vorgezeigten Gescheibe durchaus des glazialen Dwyka-tillite von Südafrika ähnlich sehen." Geheimrath Albrect Penck (Berlin) "bemerkt gleichfalls dass das von Herrn Coleman vorgelegte Gestein ganz und gar dem Tillite von Südafrika gleicht, und dass die vorgelegten gekritzten Gescheibe die typischen Merkmale glazialer Schrammung und nicht bloss pseudoglazialer Reibung tragen." Professor Grenville Cole (Dublin) "fully recognized the glacial origin of the deposit . . . especially when compared with the African Dwyka series."

Many other glacial geologists have personally expressed similar views, and one may conclude that the glacial origin of the Cobalt tillite is as well established as that of any later ice age.

The finding of slate with seasonal banding, as mentioned by Barrell,[4] who visited the Cobalt region with other members of the Geological Congress of 1913; and the pres-- ence in the slate of an occasional boulder dropped by floating ice are additional proofs of glacial conditions.

Boulder conglomerates of the Cobalt Series have been mapped by Collins with few interruptions from Cobalt to the original Huronian region north of Lake Huron, a distance of more than 200 miles in a direction from northeast. to southwest, and they have been traced almost continuously fifty miles farther northeast to the Opasatica region, where the striated surface was found. Two hundred and fifty miles still farther in that direction typical conglomerate with immense boulders occurs at Lake Chibougamau, and there are numerous intervening patches; so that the original sheet of tillite was at least 500 miles in length. From northwest to southeast the known breadth is about 100 miles; and many patches or considerable areas of boulder conglomerate probably of the same age are scattered over the Canadian shield at much greater distances from Cobalt.

Similar boulder conglomerates are found in the Huronian iron region south of Lake Superior in the United States; and the Gowganda Formation passes beneath Ordovician beds near Sault Ste. Marie and may extend far beyond the Palæozoic boundary.

It is of importance to note that the surface of the ancient rocks beneath the Cobalt conglomerate was a peneplain with very gentle relief. There were no mountains nor even high hills from which torrents could flow or where mountain glaciers could gather. The surface of granite, greenstone and Timiskaming sediments now being exposed by the weathering of the tillite has the same relief as the rest of the region covered by the Labrador ice sheet of the Pleistocene; and there is every reason to suppose that the Cobalt ice sheet began and spread out on low ground in the same way as the latest ice sheet of the region.

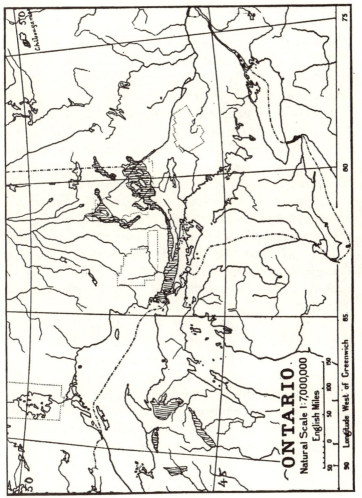

ONTARIO.
Natural Scale 1:7,000,000
English Miles

Longitude West of Greenwich

Known Area of the Cobalt Series.

Tens of thousands and probably hundreds of thousands of square miles were covered by an ice sheet of the continental type in middle latitudes—46° to 51°—and the glacial beds had about the same thickness, 500 feet, in the Cobalt region, as the Pleistocene deposits formed near the margin of the Labrador ice sheet. There is reason to suppose that the refrigeration of Huronian times in North America was almost as severe as that of the Pleistocene.

It is not to be expected, however, that the full extent of the Cobalt glaciation will ever be discovered. Most of the evidence must have been destroyed or buried out of reach during the long succeeding ages. It is indeed surprising that striated stones and striated surfaces should have been preserved at all in so ancient a formation. This can only be accounted for by the fact that the Laurentian Shield has been largely exempt from folding or faulting since it was deposited. Wherever faulting and slickensides occur, as at Timagami, forty miles south of Cobalt, striated stones are wanting, the delicate polished and striated surfaces having been destroyed.

Beneath the Gowganda tillite, north of Lake Huron, after a considerable unconformity, one finds the Bruce Series, which also includes a boulder conglomerate somewhat suggesting ice action. This lower boulder bed may be glacial, but no striated stones have yet been found in it and its origin must be considered doubtful.[5] The area of the Bruce Series is comparatively small and the conglomerate, if glacial, is of much less importance than the Gowganda tillite.

Other Tillites which May Be Huronian

The absence of fossils in Pre-cambrian rocks makes it impossible to prove that the tillites of two widely separated regions are contemporaneous. All that can be said is that they are similar lithologically and that they seem to occupy the same relative position beneath the Cambrian. Supposedly Huronian tillites have been described at a few points in the western United States.

Blackwelder has found several beds of "strongly anamorphosed boulder clays which occur in Pre-cambrian rocks on the north side of the Medicine Bow range in Wyoming. They are now biotitic schists, but they have suffered but little shearing and their relations and internal structure are clearly visible in glaciated outcrops. They are associated with finely laminated schists that suggest varve clays. The age of these beds is believed to be Lower Huronian approximately."

Hintze and Pack have described as probably of the same age as the Cobalt tillite a boulder conglomerate with a fine grained, gritty, matrix in the Cottonwood region of the Wasatch mountains and also on Stansbury Island in Great Salt Lake, Utah.[5] Subangular and facetted stones were found, but no striated ones. A thin section of the matrix has much the appearance of that of the Dwyka and of the Cobalt tillite, and specimens sent me are closely like them in every way.

No other tillites supposed to be of Huronian age have been found in America, so far as my reading goes, and the one just mentioned may really be later than the Huronian, corresponding to the late Pre-cambrian tillites described by Blackwelder from the same region.

Eero Mäkiner has mentioned briefly a probable tillite from Finland, of Kalevian age, which corresponds to the Huronian in North America, so far as one can correlate nonfossiliferous formations on opposite sides of the Atlantic. It occurs in a fragmental series including conglomerate, arkose, quartzite, mica gneiss, mica schist, calc-mica schist and dolomite. The arkose is of granite débris and unweathered, and of the boulder beds he says, "In their massive structure and low degree of assortment they bear resemblance to morainic conglomerates." They are extensively distributed, have a thickness of thirty or forty meters, and have been invaded by amphibolite and granite.

A photograph of the conglomerate by Sederholm shows rounded and subangular boulders, rather crowded together and quite like part of a moraine. No reference is made to striated stones.[7]

Grenville A. Cole has suggested a glacial origin for a Dalradian boulder bed from Donegal in Ireland. "Toward the close of the Dalradian periods—for the rocks probably included more than one system of strata—the 'boulder bed,' a coarse conglomerate of remarkable persistence, was found, and seems to point to glacial action. Drift ice in this case may have dropped boulders on the sea floor during a time of general glacial extension, for striated rock surfaces have not been found beneath the boulder bed. It is proper to state, moreover, that no striations have yet been found on the boulders themselves." [8]

It will be observed that floating ice instead of a glacier is assumed to have deposited the materials, since a striated floor has not been found. This conclusion is not necessarily correct, as shown in former chapters.

E. C. Andrews gives an account of a boulder conglomerate in the Broken Hill region of New South Wales which may be of Huronian age. It is in the Torrowangee Series, which he thinks is clearly Pre-cambrian "and may be compared with Coleman's Huronian tillites." [9]

This is probably the schist conglomerate referred to a number of years ago as occurring 27 miles N.N.W. of Broken Hill and as being much more highly metamorphosed than the South Australian Cambrian till described by Howchin. It contained fragments of schist, quartz, quartzite and granite, boulders of the latter rock sometimes reaching four feet in diameter.[10]

It may be that the Nullagine System of West Australia includes glacial deposits of the same age in its boulder conglomerates. Outcrops seen by myself in 1914 near Kalgoorlie were much like the Huronian conglomerate of Ontario, but no striated stones were found.

Striated stones, so far as known, are absent from these boulder conglomerates of the older Pre-cambrian, and none of them covers any important area. It is the widespread tillites found on the Archæan peneplain of Canada, in some places perfectly preserved in spite of their age, which demonstrate the glacial character of the period, and but for them the other occurrences would hardly be considered important. They help, however, to show the world-wide extent of the refrigeration, which must rank with the great depressions of temperature recorded in geological history.

These very ancient rocks have suffered so much destruction and metamorphism in all parts of the world that it is surprising to find so much evidence still available in regard to glaciation.

It is probable that when sought for similar boulder conglomerates will be found in other Pre-cambrian regions. For instance, a large erratic from the Permo-carboniferous of Brazil, observed at Villa Raffard near Capivary and previously mentioned, contained a variety of Archæan stones, including banded jasper, and looked exactly like some of the Cobalt tillites. Its source is unknown, but when found, the rock may extend the evidence for old Pre-cambrian glaciation to South America.

Timiskamian or Sudburian Boulder Conglomerates

Although the Huronian is the lowest series of rocks in which well preserved tillites are known, there are much older rocks which include boulder conglomerates of a very glacial appearance, though, in the nature of things, the evidence is less satisfactory in such ancient and disturbed deposits. In Canada such boulder conglomerates are found in the Timiskaming Series and the Seine Series.

The Timiskaming rocks near the lake from which the name is derived and also in the Porcupine and Kirkland Lake gold regions include boulder beds like those described from the Huronian of Cobalt, but usually steeply tilted and often squeezed so that the stones have lenticular shapes in schist conglomerates.

Examples occurring 900 feet underground in the Kirkland lake gold mine have recently been studied by Professor A. MacLean of Toronto, who suggests that they may be glacial. The pebbles and boulders, which are of Keewatin greenstone, are subangular or rounded, and one or two show indistinct striations. The matrix is of very basic materials, much more so than the matrix of most tillites; but its composition is about the same as that of the associated stratified graywacke, so that one might suppose the "rock flour" of both to have been ground from a surface of the Keewatin basic lava found in the locality.

The appearance of the specimens and of the underground photographs of a large, faintly striated boulder is decidedly glacial and it may be that unquestionable proof of ice work will be found in these rocks in spite of the fact that they have been thrust into mountain folds, penetrated by granite and profoundly eroded to form the peneplain on which the Cobalt tillite was deposited millions of years later.

In the Larder Lake region to the north, where the Timiskaming Series is well displayed, H. C. Cooke has studied the conglomerate and its associated graywackes grading into slates. The graywackes contain many subangular or angular grains of feldspar, ranging in diameter from a half to a

Probable Tillite, Timiskaming Series, West Dome Mine, Porcupine, Ont.

tenth of a millimeter; and some of the pebbles in the conglomerate, especially the smaller ones, are subangular or
sharply angular, points which would accord with glaciation;
but he thinks the deposits are of continental or subaerial
origin, formed by torrential streams in a mountain region.
The "spherical and broadly oval shapes of the larger pebbles are indicative of long continued wear before deposi-
tion. . . . The bedding, however, is hardly so variable as
in alluvial fans." He believes that the finer grained graywackes and slates were "laid down in a body of quiet,
standing water." [11]

Some years earlier the same writer described, under the
name of the Kiask Series, very similar boulder conglomerates of probably the same age at Matachawan toward the
southwest; [12] and the series is well displayed at Porcupine
to the northwest, where larger boulders, sometimes subangular, occur. These coarse conglomerates associated with
finer stratified materials, sometimes beautifully banded,
are widely distributed over an area of fifty miles by one
hundred.

Conglomerates very like them and of the same age are
found in the Sudbury Series a hundred miles to the southwest, and sediments of the same kind occur much farther
to the northwest on the Manitoba boundary. E. M. Burwash
has described the Wanipigow conglomerate from this region
and states that while no striated pebbles or other indubitable
evidences of a glacial origin were found in them, they present features that are not irreconcilable with such a theory,
since they show no bedding, and the larger fragments, which
are rounded, angular and subangular, are seldom in contact,
but generally somewhat separated in the matrix, "which
would not be the case where stream-deposited boulders or
talus blocks have their interstices afterwards washed full of
finer materials." The Wanipigow rocks were believed by

Seasonal Banding (Varves) Sudbury Series, Sudbury, Ont.

W. G. Miller to be of the same age as the Timiskaming Series.[12]

The Seine Series west of Lake Superior and far to the south of the Wanipigow region is probably of about the same age and includes boulder beds of the same kind. The deposits were described by A. C. Lawson, who studied them many years ago, as most like "fanglomerates" of desert origin. On Shoal lake this conglomerate encloses boulders of graywacke, schist, granite, felsite and banded iron formation, some of them reaching two feet in diameter, and has a glacial look, though no striated stones have been found among them.[14]

These bouldery rocks extending for 600 or 700 miles from east to west and 100 from north to south are undoubtedly land deposits and have many of the features that belong to moraines; but they have been so much disturbed that they are now in many places foliated and it can hardly be hoped that striated stones, if ever present in them, will be discovered, unless some relatively undisturbed area of them should be found.

Seasonal Banding in Timiskamian or Sudburian Rocks

Varves, or slate having bandings of a seasonal character, have been described in connection with many of the tillites mentioned in previous pages; and they may be considered evidence of the thawy summers and cold winters near the edge of an ice sheet. Similar banded deposits are found unconnected with tillite and may be taken as proofs of an important change of seasons, perhaps, however, from wet to dry instead of from warm to cold.

An example of the kind is found in the McKim graywacke at Sudbury, Ontario. The thickness of the sheets usually runs from a half inch to two or three inches, but there are sometimes two or three feet in which no stratification can be seen. The layers have a coarser part, of graywacke including angular grains of quartz and feldspar, and a finer part consisting of slate. At one place 867 layers, representing as many years, were counted in 123 feet; and the whole formation is 4,000 feet thick. Although no tillite is known near the graywacke, boulder conglomerates of the Sudbury Series

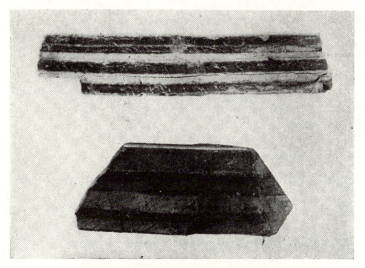

Varves, Pleistocene from Toronto, Above; Sudbury Series, Sudbury, Below.

occur not many miles away, suggesting possible ice action.[15] The unweathered feldspar grains of the coarser parts are best accounted for by a cold and moist climate.

Banded slates or schists occur at various places in the ancient rocks farther west, as at Kenora, on Lake-of-the-Woods, where there are more than 1,000 feet of them standing nearly vertical, with bands from a half inch to three or four inches thick.

Sederholm has described and figured similarly banded rocks in the Bothnian of Finland, which is probably of about the same age, and he interprets them as implying a cold climate with snowy winters.[16]

The Doré Conglomerate

The still older Doré conglomerate has attracted the attention of geologists from the days of Logan, who placed it in the Huronian. The latest geologist to study it is Collins, Director of the Canadian Geological Survey, who has been good enough to allow me to read the portion of a still unpublished memoir which refers to it. He finds that the conglomerate lies between lower and upper volcanics of the Keewatin, so that it is of Keewatin age. The greatest thickness of the largest band or lens is given as 8,400 feet, but it diminishes in each direction rapidly, and there are smaller but still important bands not connected with it. The dip varies from 50° to the vertical.

There is a gradual change from conglomerate and graywacke of distinctly sedimentary appearance into materials which resemble scree or coarse or fine tuffs.

It is a land formation with no sharply erosional base and varies enormously in thickness within short distances. The series contains volcanic rocks apparently interstratified with the sediments. There is little assortment of the materials and the original feldspars are fresh.

In the manuscript there is no definite statement as to the origin of the rock, but apparently it is looked upon as composed of materials from nearby mountains. The possibility of ice action is not mentioned, and in conversation Dr. Collins doubted the probability that the conglomerate was formed by glaciers.

It is evident that a conglomerate which is in places 8,000 feet thick and which extends, with some interruptions, for thirty or forty miles, must have required forces operating on a large scale.

Photographed by W. H. Collins.

Possible Tillite, Doré Conglomerate, Keewatin, Michipicoten, Lake Superior.

In my own field work certain islands near the mouth of Doré River were found to present on glaciated surfaces the best preserved portions of the conglomerate, which may be briefly described. The stones were of all sizes from a foot or two in diameter to the smallest pebbles, and the matrix consisted of the same materials as the enclosed stones down to particles forming arkose. In a square yard of surface eighty-one stones were counted, including green schist, granite, iron range rock of granular silica looking like sandstone, felsite, etc., the first three rocks being commonest. Most of the stones were fairly well rounded and few were angular. The largest boulder seen was of granite two feet in diameter. The nearest known outcrop of granite was more than two miles away.

Most of the conglomerate is a good deal squeezed, and the boulders of schist are often rolled out into lenses, so that it was not possible to extract any uncrushed stones except the more resistant granites, and none of them showed signs of glaciation.[17]

The modes by which such coarse materials, derived from several different kinds of rocks, could be piled up to a thickness of 8,400 feet and spread over an area thirty or forty miles long, with an unknown width, are very few. Mountain torrents might accomplish it if the mountains were high enough and near enough.

Possibly extensive mud flows in a mountainous region might do the work. The conglomerate must have been formed by the aid of either water or ice, since many of the boulders are well rounded and could not have resulted from talus deposits. It seems to me that these heterogeneous boulder conglomerates, with unweathered feldspars in their matrix, piled up to a thickness of thousands of feet, are more naturally explained by the action of mountain glaciers and glacial streams than in any other way.

As the Keewatin is one of the oldest known of the world's formations, the probability of glaciers having been at work on mountain slopes at the time is of considerable interest. It is unlikely, however, that anything more than a probability of glaciation will ever be reached in regard to a rock so ancient and which has undergone mountain-building stresses as extensive as those shown in the Keewatin.

Darwar Boulder Conglomerate in India

The Kaldrug conglomerates of the Pre-cambrian Darwars in the gold region of southern India, as described by Foote, are very like the most ancient boulder conglomerates of Canada and suggest the same origin. There are "great out-crops of an extraordinary conglomerate of extreme coarse-ness. The pebbles, often approaching in size to small boulders, consist of granite, generally cemented together in a foliated chloritic matrix." The conglomerates are associated with quartzite.[18]

Conclusions

From the foregoing account of Pre-cambrian boulder conglomerates it is seen that ice action has been proved or shown to be probable in the Keweenawan, the Animikie, the Huronian, the Timiskaming and the Keewatin, i.e., in all of the main subdivisions récognized in Canada, where the largest area of Pre-cambrian rocks in the world is exposed. In the case of the Huronian (Cobalt or Gowganda) the area of glaciation is so great that one must assume a very im-portant refrigeration, one to be classed among the first four in severity.

In other continents, especially Africa and probably Aus-tralia, the late Pre-cambrian ice sheets also were of great ex-tent and even reached what are now warm temperate regions.

When one considers the age of the Pre-cambrian rocks and all the possibilities of the destruction or permanent burial or submergence of the tillite, or else of its complete metamorphism into schists, the amount of evidence of glacial work is very surprising.

Not very long ago the ancient gneisses and schists were looked on as remnants of the original crust of the earth, formed when it cooled from the molten condition demanded by the Nebular Hypothesis. Now it appears probable that the Pre-cambrian was the coldest part of the earth's history, with glaciers at work within every one of its main subdivisions. Part of this effect may, however, be due to perspective, the more distant glaciations being apparently crowded together because we do not appreciate the intervals between them.

The finding of these evidences of ancient glaciation has revolutionized our ideas as to the origin and history of the world and has rendered improbable the theory which was accepted by both astronomers and geologists a generation ago as the natural starting point in the evolution of the planet.

REFERENCES

1. "A Lower Huronian Ice Age," *Am. Jour. Sc.,* Vol. XXIII, 1907, pp. 187-192; also *Jour. Geol.,* Vol. XVI, 1908, pp. 149-158.
2. *Bur. Mines, Ont.,* 1918, Part 1, p. 228.
3. *Geol. Sur. Can., Sum. Rep.,* 1922, Part D, pp. 63-4.
4. "Discussion of Varves," *Bull. Geol. Soc. Am.,* Vol. 27, 1915, pp. 112-113.
5. COLLINS, *Geol. Sur. Can., Mus. Bull.,* No. 8; also QUIRKE, *Sum. Rep.* 1920, Part D, p. 12.
6. *An. New York Acad. Sc.,* Vol. XXII, pp. 99-101.
7. *Bull. Com. Geologique de Finlande,* No. 47, pp. 20, 21 and 148.
8. *Geology of Ireland,* Sonderdruck aus dem "Handbuch der Regionalen Geologie," III Band. 1 Abt. p. 44.
9. *Am. Jour. Sc.,* Vol. IV, 1922, pp. 504-5, a "Review of the Geology of Broken Hill District;" *Mem. Geol. Sur. N.S.W.* No. 8.
10. *Austr. Ass. Adv. Sc.,* Vol. XII. 1909, p. 238.

11. *Geol. Sur. Can.,* Mem. 131, pp. 25-42.
12. *Ibid.,* Mem. 115, pp. 20, 28 and 29.
13. *Bur. Mines, Ont.,* Vol. XXXII, Part II, 1923, p. 29.
14. *Ibid.,* Vol. V. 1895, pp. 95-6.
15. *Problems of Am. Geol.,* Yale Univ., 1913, pp. 93-4.
16. *Com. Geologique, Finlande,* No. 24, pp. 22-3.
17. *Bur. Mines, Ont.,* Vol. XI, 1902, pp. 163-7.
18. *Geol. Sur. Ind. Recs.,* Vol. XXI, Part II, p. 48.

PART III

CAUSES OF GLACIATION

Introduction

Features that Have to Be Accounted for

FROM the preceding study of the known ice ages certain general conclusions may be drawn to serve as tests of the validity of any theory which attempts to account for such climatic changes.

Under normal conditions the world has a relatively mild and equable climate with no permanent ice at low levels even in the polar regions.

From time to time, though at irregular intervals throughout its history, there have been relatively short periods of cold accompanied by a great extension of mountain glaciers and sometimes also by the formation of ice sheets at low levels. In the most severe visitation of the kind ice sheets invaded the tropics on three or perhaps four continents.

Ice ages are, in most cases, broken by interglacial periods of milder climate. Sometimes this occurs two or three or more times, indicating a comparatively rapid oscillation from cold to warm and warm to cold.

All parts of the world have their temperature lowered during an ice age, the tropics as well as the temperate and arctic zones.

The earth has not been gradually cooling from age to age. Liquid water has existed upon it as far back as the geological record goes. Since life appeared the water of the earth has never been wholly frozen nor wholly vaporized, which implies a limitation of the extremes of temperature. Some part of the earth has always been above the freezing point in the coldest times, and some portion has always been at a temperature below 190° Fahr. These limitations have existed for hundreds of millions of years.

Types of Theories Proposed to Account for Ice Ages

Scores of methods of accounting for ice ages have been proposed, and probably no other geological problem has been so earnestly discussed, not only by geologists but by meteorologists and biologists; and yet no theory is generally accepted. The opinions of those who have written on the subject are hopelessly in contradiction with one another, and good authorities are arrayed on opposite sides, so that in the maze of suggestions and counter suggestions it is very difficult to present the matter fully and fairly; yet the subject is of such tremendous importance in the history of life and of the physical development of the world that it must be discussed in a work of the present kind.

In general the theories intended to explain the causes of ice ages may be classed as geological, meteorological or astronomic. In many cases, however, causes of more than one sort are combined, so that a strict separation is not possible.

The geological theories have to do with the secular cooling of the earth, elevations and depressions of the earth's crust, the changing relations of land and sea, the shifting of continents and of ocean currents.

Meteorological theories bring in the direction of storm paths, the amount of evaporation of water, the presence of dust in the air and the proportions of the different gases making up the atmosphere.

Astronomical theories turn on the shape of the earth's orbit, the obliquity of its axis, the variability of the sun's heat, etc.

It is intended to outline and discuss the principal theories that have been proposed, and to suggest their probable importance in solving the problems of glacial periods.

It will be understood that an immense literature has grown up on this subject, beginning more than half a century ago and expanding ever since. Many of the earlier discussions have at present only an interest of an historical kind, since later discoveries have shaken or destroyed the fundamental assumptions on which' they were based; but certain points of view that were brought forward long ago are still being discussed.

Early works on the Glacial Period, such as James Geikie's and F. G. Wright's, are well worthy of study still; and the elaborate books of Croll on "Climate and Time" (1875) and "Climate and Cosmology" (1886), still repay perusal. W. B. Wright's book on the "Quaternary Ice Age" is excellent also. Several of the modern textbooks of geology discuss glacial matters very instructively; such as Chamberlin and Salisbury's Geology, which is of great interest.

Quite recently, two books have appeared suggesting new and entirely different explanations of glaciation, Humphrey's "Physics of the Air," (1920) and Huntington and Visher's "Climatic Changes" (1922). Both are well worth reading. The former makes volcanic dust the most important factor, and the latter find the essential cause of climatic changes in the variation of the sun's radiation.

Prof. Hobbs' "Characteristics of Existing Glaciers" (1911), and Wright and Priestley's "Glaciology of Antarctica" (1922) give very useful accounts of modern glaciers and ice sheets, which serve as models of the machinery which worked so much more widely in the great ice ages.

Smaller articles on the subject are innumerable; and in

the preparation of this work more than two hundred books and articles have been consulted.

Sir Edgeworth David and also Prof. Schuchert have given excellent accounts of ancient glaciation.

CHAPTER XIV

The Supposed Cooling of the Earth

WHEN Agassiz proved the former great extension of glaciers and ice sheets the idea was received at first with incredulity, but later was accepted and adjusted to the prevalent theory of a slowly cooling earth which began as a lens of intensely hot gas, became a molten sphere with an enormous atmosphere, acquired a solid crust in whose depressions water could condense, though still at the boiling point, and finally reached a temperature endurable for living beings. Then geology began, but the gradual cooling of earth and sun continued through the hot Palæozoic times and the more temperate Mesozoic, and at the close of the Cenozoic reached the intense refrigeration of the Glacial Period. The hopeless future that awaited us at no distant time was that of a sun grown cold and all life extinct on a frigid world after the pattern of the moon.

This melancholy theory was rendered untenable, at least in its usual form, by the undoubted proof of an even more severe visitation many millions of years earlier, at the end of the Palæozoic, from which the earth completely recovered; yet so simple and long accepted an account of the world's beginning does not pass without a struggle, and even now there are textbooks of historical geology that commence in the orthodox way with a hot earth. This is more frequent in Europe than in America.

In America an ingeniously modified theory of a cooling

earth has been worked out somewhat elaborately by Marsden Manson and is accepted by some geologists and palæobotanists. Manson has published a number of interesting papers elaborating the view that the mild climates so prevalent in the earlier history of the world were due to its internal heat which gradually diminished, though largely retained until the end of the Pliocene by a sheltering mantle of clouds, which also prevented solar heat from penetrating to the earth's surface. Owing to the final cooling of the earth, so that the ocean was chilled and evaporation was greatly reduced, the cloud cover partly disappeared in the Pleistocene and the earth's climates came under solar control.

In his latest publication, "The Evolution of Climates," he introduces some new ideas, the most important being the intermittent supply of earth heat to the ocean by periodic changes in the form of the earth which made increments of energy available. At long intervals heat was set free by "crustal ruptures, denudation, exposure of radio-active substances, chemical and physical actions and reactions, etc."

His point of view may be stated in his own words, as follows: "Earth heat was not available to prevent glaciation when the intervals between its liberation from the crust were too long, but as stored heat in the more slowly cooling oceans it was available to maintain water vapor for a denser cloudiness than now prevails. For the foregoing reasons, and others more specifically given later, it is held that earth heat was an effective climatic factor during all of geologic time, that its effects varied in intensity, notably upon continents; and that glaciation was caused by the interception of solar radiation by clouds maintained by successive increments of ocean-stored earth heat during intervals between its liberation from the non-conducting crust long enough to permit land areas to cool below freezing; and that no glaciation under solar control of climates has been recorded." [3]

The oceans have now cooled so far that only 52 per cent of the sky is cloudy, permitting the sun's heat to control climates and introducing our present arrangement of zones. If there are any future glacial periods they must arise in a different way from those of the past.

He assumes a very sensitive balance of relations between the temperature of the more easily cooled land and that of the sea, which was more constant because of the high specific heat of water. Thus at the time of the Permo-carboniferous glaciation the tropical lands were cooled below the freezing point, though the ocean retained enough heat to keep up the mantle of clouds by evaporation, preventing the sun from warming up the equatorial and tropical latitudes as it does now. At that time the latitudes between 15° and 35° north and south were exposed to cold anticyclone winds which reached the surface and produced ice sheets, while the temperate and arctic zones were exempt. A slight fall in the temperature of the sea would allow the clouds to thin in the tropical belt and solar heat would remove the ice. The melting of the ice would remove a load from the land and cause crustal disturbances by which the sea would be slightly warmed, renewing the cloud covering and causing fresh glaciation. Thus interglacial periods are accounted for. A more profound set of crustal disturbances at the end of the Palæozoic ice age warmed up the sea to a greater degree and the world's climates became uniformly mild once more.

Apparently he admits that there were times of partial solar control in the interglacial periods of the Permo-carboniferous, but during the rest of geological time up to the Pleistocene the covering of clouds was unbroken. The other glaciations he accounts for as due to elevation of the regions where ice was formed.

The Mansonian theory has been strongly supported by Dr. F. K. Knowlton, a well-known palæobotanist,[4] who gives an admirable summary of the development of plants in the world during the successive geological ages, and emphasizes the evidence of mild or even tropical climates extending into temperate and often arctic regions during most if not all of the subdivisions of geological time. Most of his ancient floras suggest moist and equable conditions with little or no change of seasons, as shown by the usual absence of annual rings in the wood and by the tropical or subtropical character of their nearest modern relatives. The array is very impressive and makes an excellent introduction to a theory of ancient climates implying hothouse conditions. He reinforces the evidence of the fossil plants with that of fossil animals which often show a world-wide uniformity in the temperature of the sea.

It is admitted by all geologists that during by far the greater part of its history the earth has enjoyed mild climates with little evidence of zones, and that in some periods even the north and south polar regions have had warm temperate conditions. The organic remains enclosed in the rocks make this certain; but, as shown in earlier chapters of this work, there have been a number of comparatively short periods of cold scattered through geological time, critical periods of immense importance in the development of the life of the world.

The evidence of the fossil plants is very partial, since plants thrive during mild and moist conditions and are then entombed and preserved, while they are few and impoverished in times of cold and drought, and conditions are then unfavorable for their preservation. The great floras of the ancient world come mostly from the periods when coal was deposited, which of necessity were times of relatively mild temperature and of sufficient moisture for rich plant growth.

The material he works with necessarily gives a bias to the

mind of the palæobotanist. The evidence of moisture and mildness is overwhelmingly strong, while the times of cold and dryness are poorly or not at all recorded in his fossils.

Much of the same is true of palæozoology, since glacial and arid conditions belong to the land and may have little effect on the sea. When land deposits are being formed there is usually no record of the life of the time. The lands are then in emergence and the shallow epicontinental seas which entomb most of the fossils are greatly diminished.

It will be seen then that the evidence afforded by plants on the land and by animals in the sea will be partial to the mild and moist periods. For proofs of glaciation and also of arid conditions our reliance must be upon the physical characters of the rocks formed on the land by ice sheets or by desert winds.[5]

The facts brought out in an earlier chapter in regard to a widespread late Pre-cambrian glaciation and an even more striking glaciation on low ground in the Huronian, along with the strong probability that still older boulder conglomerates were caused by ice, prove that the earth was not warmer, on the surface at least, in those early times than in later ages, although, according to the Mansonian theory, the earth as a whole has been losing heat ever since.

There are other lines of evidence also that strongly oppose the supposition of a continuous screen of clouds. Although Knowlton finds few examples of annual rings in ancient trees there are annual layers, due to change of seasons, either from warm to cold or from wet to dry, in sedimentary rocks of almost all ages. The varves formed in the lakes near the ends of retreating glaciers are found not only in the Pleistocene but also in the Permo-carboniferous and the Huronian; and shales or slates or graywackes with annual layers are known from a number of the older formations, such as the McKim graywacke of the

Sudbury region with layers from a half inch to two or three inches thick, the series having a total thickness of 4,000 feet.[6] This belongs to the earlier Pre-cambrian; and Sederholm reports a similar example of annual layers from the Bothnian of Finland.[7]

One cannot imagine any other rhythm so uniform and continuous for hundreds or thousands of years as that of the changes of season due to the inclination of the earth's axis to the plane of the ecliptic; and these changes imply a solar control.

The evidences of drought, also, cannot be accounted for in a world continuously covered with clouds and enveloped in a moist heat. Desert formations are known from the Triassic, the Permian, the Devonian, the Silurian and the late Pre-cambrian. They include not only wind-blown sands but mud cracks due to drying and great deposits of various salts where inland seas have been evaporated. Some of these deposits are found much farther from the equator than our present desert zones, indicating a more intense aridity than that existing now.

There is no conceivable cause for these deposits except strong sunshine and they seem quite incompatible with a cloud-enveloped earth.

Although the theory so skillfully worked out by Manson and supported by Knowlton is attractive as conforming to a view long held by geologists, that the earth was once molten and has been cooling down since the beginning, the advancing knowledge of glacial and desert formations reaching far back toward the beginning of known geological time makes it unacceptable. If there ever was a hot earth enclosed by unbroken clouds it seems to have cooled to conditions not unlike those of the Pleistocene before the Huronian, and so has little bearing on geological history.

Theories Connected with Changes of Level

It is well known that the crust of the earth has undergone great changes of level, sea bottoms becoming land surfaces and marine rocks being raised as mountain chains in some parts of the world, while blocks of the crust seem to have sunk beneath the sea in others. It is also well known that the temperature is lowered about 1° Fahr. for every 300 feet of elevation. At a sufficient height all over the world snowline is reached, above which snow (neve) accumulates from year to year and drains downward as glaciers.

Why not account for glaciation by raising the area above the level of perpetual snow? This very natural method of providing for glaciation has been proposed by a number of geologists, some of them having long experience in Pleistocene field work, such as G. F. Wright [8] and Warren Upham; and it has been suggested that continental ice sheets are formed on tablelands and not on low ground.[9] It has been stated that the Labrador ice sheet took its rise on the Laurentian *plateau*. As a fact it began in the lower parts of the Laurentian Shield and later spread out over the higher, more plateau-like edges of the shield.

There are, however, some proofs that at least a portion of the area covered by the Labrador ice sheet was higher in pre-glacial times than now. Before the ice age the region of the present Great Lakes and the St. Lawrence stood about 2,000 feet higher than now, as shown by an old channel carved to that depth beneath the sea at the edge of the continental platform between Nova Scotia and Newfoundland.

An elevation of 2,000 feet would give a colder climate but would be far from reaching snowline, so that this alone would not cause glaciation. It might be assumed, of course, that the elevation was really much greater than that suggested by the old channel, but there is no proof of this.

When one considers the distribution of ice sheets in the Pleistocene, covering 4,000,000 square miles of North America and half as much of Europe, as well as Greenland, Iceland and Spitzbergen; and also the greatly lengthened glaciers of the Rocky Mountains, the Alps, the Himalayas, the Andes, the Atlas Mountains, Ruwenzori and its fellow peaks in central Africa, Kosiusko in Australia, the southern island of New Zealand, and Patagonia in South America, it becomes evident that all parts of the world could not have been elevated at once. The theory breaks down of its own weight. Besides, it is possible that in some important regions a lowering of the land would increase glaciation. It has been suggested that this would be the case in the Antarctic continent.

It cannot be doubted that elevation above snowline would cause local glaciation, but there is no evidence that large scale ice sheets can be formed in this way, and that a universal refrigeration, like that of the Pleistocene, could be produced thus is manifestly impossible.

Nevertheless relative levels have a very important bearing in determining the special areas which will be glaciated if the climate becomes cold. The Yukon territory and inner Alaska were not ice covered in the Pleistocene because protected by lofty mountains from the moist winds of the Pacific, and Siberia bore no large ice sheet because great mountain ranges rose between it and the warm seas.

The Effects of Depression

On the other hand depression of land barriers, by giving new routes for ocean currents, warm or cold, may greatly influence glaciation.

It has sometimes been suggested that lowering the Isthmus

of Panama a few hundred feet, allowing the Gulf Stream to flow on to the west instead of its being deflected northeast toward Europe, might produce glaciation in the north of that continent. Just what would happen in that event is not easy to predict, though probably western Europe would lose its mild climate and conform to the temperatures appropriate to its latitude.

It will be understood that such effects as have been mentioned could only occur at times when the polar regions were ice covered. When the climate was universally mild such deflections of ocean currents would have little effect, and probably the uniformity of climate would greatly diminish the force of the prevalent winds, and in this way remove the main driving power of the ocean currents. Mild climate to the poles with only slight differences of temperature between the arctic and the tropic zones would imply also a languid circulation of the atmosphere and the ocean. It would be a time of relative stagnation.

The Supposed Drift of Continents

One of the most eagerly discussed geological theories having a bearing on glaciation is that of the movement or drift of land masses from one place to another on the surface of the earth. So far as I am aware the first suggestion of the shifting of continental masses was made in 1909 by F. B. Taylor, who called attention to the peculiar relations of northern Canada, Greenland and Scandinavia, which look as if they might be fitted together. He supposes that North America drifted to the southwest, thrusting up the chain of the Rocky Mountains on its way.[10]

This he did not associate with glacial phenomena; but a later suggestion by Wegener, of former relations between India, South Africa, Australia and South America in a supposed continent of Gondwanaland, has been connected with the Permo-carboniferous glaciation and deserves special mention.[11]

The name Gondwanaland comes from the Gondwana Series of India, having as its lowest member the Talchirs, whose glacial beds have already been described; and the supposed continent was put forward originally to account for the distribution of the *Gangamopteris* and *Glossopteris* flora which followed up the Permo-carboniferous glaciation, and other features which India, South Africa and Australia had in common. At first the suggestion was applied only to the three glaciated regions around the Indian Ocean; but the finding of glacial beds and of the *Gangamopteris* flora in South America caused this continent also to be included.

It was thought that the spread of the peculiar cold-climate ferns and also of some of the animals which came after the ice age could only be accounted for by land connections between these now widely separated regions. Hence the supposition of a vast continent spanning the Indian Ocean. The powerful glaciation of all of these regions was by some connected with the conception of a great land mass where there is now deep sea, and elevation was looked on as a cause of the glaciation.

Wegener is sponsor for a different suggestion in regard to Gondwanaland, in which all the glaciated regions, India, South Africa, Australia, South America and Antarctica, are combined as a single vast land mass; which later broke up, the constituent parts drifting asunder and taking up their present positions.

Du Toit in South Africa, one of the most careful students of the Dwyka, eagerly accepts this theory and supports it as a method of accounting for the powerful glaciation of his own region in the Permo-carboniferous.[12] In his exceedingly interesting paper on The Carboniferous Glaciation of South Africa, he shows the supposed continent on a sketch map, in which these land masses are brought close together, the prominence of Brazil fitting into the Gulf of Guinea, and the tip of the Indian peninsula being brought down opposite the end of South Africa. The glaciated area is shown with a length of 8,000 miles and a breadth of 6,000; but a long gulf is represented as reaching between India and Australia and ending near the southern point of Africa. Antarctica has been drawn toward India so that the south pole stands close to the outer edge of the continent.

Gondwanaland as a whole is made twice as large as the glaciated area, and includes India, most of Africa, two-thirds of South America, Antarctica, and three-quarters of Australia. As shown in the sketch it would be larger than Eurasia at present.

Du Toit's sketch map differs in important ways from Wegener's; particularly in the position assigned to the south pole, which Wegener places near the south end of Africa. As arranged on the map all the glaciated areas come within the 45th parallel, south, which he compares with the Pleistocene limits of lat. 52° in Europe and 37½° in America. Unlike some others, who have advocated a Gondwanaland of another type, he does not suggest that the glaciation was due to elevation, but thinks that the land nowhere stood much higher than a few thousands of feet above sea level. In several continents the ice reached sea level.

If there was such a Permo-carboniferous continent in the southern hemisphere within 45° of the pole, many difficulties would be removed in accounting for the known glaciation, though the vast area of the supposed continent would be hostile to a continuous glaciation over the whole area.

Ice sheets can be formed only when moist winds can reach the glaciated region before all their snow has been deposited.

One reason why Siberia was not ice covered in the Pleistocene was its distance from warm seas. Cold seas admit of too little evaporation to provide the requisite precipitation. The immense diameters of the supposedly ice covered Gondwanaland, 8,000 miles by 6,000, make the supply of snow hard to account for.

During the Pleistocene ice age the center of the ice sheet most distant from the sea, that of the Keewatin area, was only 1,100 miles east of the Pacific, 1,850 miles north of the Gulf of Mexico and 1,500 miles west of the Atlantic. Sir Frederick Stupart, Director of the Meteorologic Service of Canada, believes that most of the snow supply of the Keewatin and Labrador ice sheets came from the Gulf of Mexico, which was less than 2,000 miles away from their respective centers.

The supply of moisture for the vast Permo-carboniferous ice sheet, as shown on Du Toit's map, would have to come from more than double the distance mentioned above. This seems to me a real difficulty if one admits that the glaciated regions were grouped together in the way suggested; but, after all, the main doubt one feels in the matter is in regard to the physical possibility of continental masses wandering apart for thousands of miles in opposite directions. In the case of India there would have to be a voyage of 3,500 miles to the north, while South America would travel 2,000 miles west, Australia 1,000 or more miles east, and Antarctica 1,500 miles south.

The theory of isostasy becomes more and more certain as gravity determinations are extended over the continents. The variations from the theoretical balance of the land surfaces prove to be comparatively small; and one may say that the earth's crust, so far as examined, is practically in isostatic equilibrium. This means, of course, that if a continent floated to a new position, leaving a deep ocean in its place, there must be an equal return flow of deep-seated heavier matter to replace the mass removed.

When the continental masses making up Gondwanaland scattered in all directions, leaving the Indian and South Atlantic oceans, averaging more than 10,000 feet in depth, in the positions they had occupied, the thickness of the earth's crust was diminished to the extent of at least two miles. If isostatic equilibrium was to be maintained there must have been a deep-seated migration of an equal weight of heavy magma to restore the balance.

One must think of the continents as rafts of light rock supported at an appropriate level by a heavier magma. The sea bottoms will consist, as has been suggested by Daly and others, of heavy rock, such as basalt, which floats at a correspondingly lower level. The yielding magma which supports both will be forty miles or more below the surface. A continent may be thought of as a monstrous iceberg with much the largest part of its bulk submerged in the underlying plastic material.

The usual rocks found on continents, the sediments, granite, gneiss, etc., have an average specific gravity of 2.6 or 2.7; while the basic rocks supposed to make the sea bottoms would have a greater density, perhaps 2.9 or 3, and would be several per cent heavier.

The continents float on the average 2½ miles higher than the sea bottom, and the light rocks causing their elevation must extend downwards far enough to support this load.

To permit the migration of the continents we must suppose these rafts to plough through the solid sea-bottom crust while an equivalent mass of heavier magma moves the same distance in the opposite direction. There must always be a counterbalancing return of heavier underlying material.

Even allowing for an extremely slow drift of the continents since Triassic times, a shift of 3,000 miles under these conditions, as in the case of India, seems incredible; and one can conceive of no possible force that would propel the separating continents in different directions. The change in the bulge of the equator due to a slowing down of the earth's rotation has been suggested as a possible cause for such continental drift; but this cause can hardly account for the movement of land masses in all directions at once.

If the motions had been from the pole along different meridians toward the equator an explanation would not seem so difficult, but Du Toit points out that the glaciated areas were eccentric as regards the south pole, just as the Pleistocene glaciation was eccentric with reference to the north pole. Moreover, the Indian peninsula, according to the supposition, moved not only toward the equator but passed several degrees beyond it.

The distribution of the cold climate ferns and of certain amphibia following the glaciation would be naturally accounted for if the lands on which they occurred had at one time been connected, but the connection may have been much more roundabout. The spores of the ferns might easily be distributed far and wide by the winds. They probably originated in Antarctica, which may have been joined by land or by chains of islands to the southern continents, Africa, Australia, and South America; and India has at numerous times been connected with Africa. Some such mode of distribution seems more natural, and also more economical of energy, than the assumption that all these regions once coalesced as a single continent and then drifted asunder.

The subject of the distribution of species in geological time has been discussed by W. D. Matthew in a very illuminating way; and he shows that there are many modes of accounting for the known distribution otherwise than by the wholesale construction of continents where there are now deep oceans. His study of the subject was made before the drifting of continents had been suggested, but his conclusions seem to make quite unnecessary the incredible theories now attracting attention.[13]

Wegener considers that the wandering of the earth's poles was a very important factor in the great glaciations of the Pleistocene and the Permo-carboniferous, the north pole being in the center of the assembled northern continents, in the first case, and the south pole in a similar position in the earlier glaciation. Whether the poles have migrated in the way suggested seems very doubtful. The subject of a change of position of the earth's axis as a cause of glaciation will be discussed later; but it may be suggested here that the earth is a gyroscope, and as such, has a very powerful tendency to keep its axis of rotation pointing continuously in the same direction. Any sudden change in the direction would probably wreck the world completely.

Changes in Ocean Currents

Ocean currents are closely related to the shapes and connections of the continents, to the prevalent winds, and to the earth's rotation on its axis. They affect climates powerfully and rearrangements of the land connections, changing the oceanic circulation, have been suggested as at least contributory causes of glaciation.

One of the strongest advocates of this idea was Sir William Dawson, who explained the Pleistocene glaciation of eastern North America in this way. At present there is a very striking difference in the climate on the two sides of the north Atlantic. Europe has mild winters along the whole Atlantic coast, even as far north as Hammerfest; harbors in northern Norway within the Arctic circle are open all winter; while Quebec, in lat. 47°, is closed for four or five months by ice. The whole Labrador coast has arctic conditions and only two or three months of summer, as contrasted with the mild climate of Ireland and Scotland in the same latitudes.

The cause of these astonishing differences is of course the arrangement of the ocean currents, the Gulf Stream carrying the warm water of the tropics to the north Atlantic where, as a surface drift, it washes the shores of Europe and warms the prevalent westerly winds. On the other hand an arctic current flows south from Davis Strait carrying bergs and ice floes along the coast of Labrador as far south as the banks of Newfoundland, chilling the whole coastal region.

Dawson calls attention to these facts and accounts for the boulder clay and drift deposits of eastern Canada and the states immediately to the south by supposing a lowering of the land toward Hudson Bay and the north, allowing a broader arctic current to sweep its burden of ice to the Gulf of St. Lawrence, thus lowering the temperature and permitting local glaciers to form on the mountains.

The European glaciation he explained, somewhat doubtfully, as due to a lowering of the Isthmus of Panama, allowing the warm waters of the tropical Atlantic to pass on into the Pacific instead of curving round the Gulf of Mexico to form the Gulf Stream.[14]

That changes in the level of the land, allowing the shifting of ocean currents as suggested by Dawson, would lower the temperature of Canada and northern Europe and aid in glaciation is probable; though it should be remembered that only twelve or fourteen tiny "snow-drift" glaciers are known in Labrador, in spite of the icy current sweeping from Davis Strait along its shore. The glaciers cease about thirty miles inland, and no ice sheets occur in the interior. Probably the evaporation from the arctic current, which has a temperature a little below freezing, is too slight to provide the necessary snow for glaciation. The climate, like that of Spitz-bergen, is semi-arid.[15]

However, the whole theory has dropped out of considera-tion since it has been proved that the depression south of Hudson Bay did not reach sea level, so that no arctic current traversed the interior of North America in the ice age.[16] There is also no foundation for the belief that the Isthmus of Panama was sunk below sea level during the Pleistocene. The Gulf Stream existed during the glacial period and its warm waters provided the moisture for the Scandinavian glacial center.

One may conclude that the arrangement of ocean currents powerfully affects the distribution of heat and cold on the earth, and may be an auxiliary cause of glaciation in ice ages but cannot be considered as a primary cause.

There is also a vertical circulation of the sea waters that is of importance as affecting climates. The icy water of the Arctic and Antarctic seas is heavier than warm water, settles to the bottom and invades the ocean depths even within the tropics, where deep-sea temperatures are only a degree or two above the freezing point. This, no doubt, cools the ocean floor. On the other hand, there may be a vertical circulation of an opposite kind in which evaporation in equatorial regions renders the water more saline, and there-fore heavier, so that it sinks to the bottom and spreads out in the depths. Ultimately this warm water reaches the polar regions and is forced to the surface, where it helps to produce a mild climate.

Chamberlin and Salisbury look on these two types of circulation as of considerable importance, the one as assist-ing in polar glaciation, the other supplying warmth to the Arctic and Antarctic regions.[17]

The existence of one or the other type of circulation would, no doubt, affect the climate of temperate and polar regions and would aid the work of other climatic factors; but the balance between the two methods of circulation seems close and it is not easy to predict exactly what the results of a reversal would be. In any case one must suppose that the change of climate had already reached an important stage before the changes in oceanic circulation could begin; so that the latter could not be a primary cause either of glaciation or of a mild climate, but only subsidiary.

REFERENCES

1. "Conditions of Climate in different Geological Epochs," etc., *Mex. Geol. Congr.* 1907.
2. "Climates of Geol. Time," *Carnegie Inst., Pub.* 192.
3. *The Evolution of Climates*, p. 29.
4. "The Evolution of Geological Climates," *Bull. Geol. Soc., Am.*, Vol. 30, 1919, pp. 499-566.
5. SCHUCHERT, "Evolution of Geological Climates," *Am. Jour. Sc.*, Vol. I, 1921; COLEMAN, "Dry Land in Geol.," *Bull. Geol. Soc. Am.*, 1916, pp. 175-192; COLEMAN, "Palæobotany and the Earth's Early History," *Am. Jour. Sc.*, Vol. I, 1921; SAYLES, "The Dilemma of the Palaeoclimatologists," *Ibid.*, Vol. III, 1922.
6. *Problems of American Geology*, Yale Univ., 1913, pp. 93-4.
7. "Subdivisions of Pre-cambrian of Fenno-Scandia," *Compte Rendu, Geol. Congr. II*, pp. 683-698.
8. *The Ice Age of North America.*
9. *Antarctic Glaciology*, 1910-13, p. 468.
10. "Bearing of the Tertiary Mountain Belt on the Origin of the Earth's Plan," *Geol. Soc. Am.*, 21 (2), pp. 179, etc.
11. *Die Entstehung der Kontinente und Oceane.*
12. *Trans. Geol. Soc. S. Af.*, Vol. XXIV, 1921, pp. 218-227; also *S. Af. Jour. Sc.*, Vol. XVIII, Nos. 1 and 2, pp. 120-140.
13. "Climate and Evolution," *N. Y. Acad. Sc.*, Vol. XXIV, 1915, pp. 171-318.
14. *The Canadian Ice Age*, 1893.
15. "Northeastern Labrador," *Geol. Sur. Can.*, Mem. 124, 1921, p. 8.
16. "Glacial and Post-glacial Lakes," *Univ. Toronto Stuaies, Ont., Fisheries Research Lab.*, No. 21, 1922.
17. *Geology*, Vol. III, pp. 439, etc.

CHAPTER XV

Atmospheric Changes as Causes of Glaciation

THAT the atmosphere affects climates in important ways need scarcely be mentioned. Changes in the circulation and composition of the air give us our variations of weather, and the direction of air currents may quickly bring frigid or warm conditions at a given place, as every one is aware, and a heightening of such conditions for a period of time would mean a change of climate. Atmospheric changes may be imagined as going far enough in the course of ages to transform a mild climate into a severe one allowing ice sheets to accumulate, or as reversing the process.

Changes in the usual paths of cyclones have been suggested by Harmer and Brooks as playing a great part in causing ice ages.[1,2] In this case the composition of the air is not supposed to be modified, but only its direction of circulation.

Beside the important gases nitrogen and oxygen which make up most of the atmosphere, there are present in the mixture of gases, vapor of water, which is constantly varying in amount, and carbon dioxide, which varies very slowly and is present to the extent of only three parts per ten thousand. The two permanent gases are very transparent to light and heat; vapor of water is much less transparent; and carbon dioxide is transparent to light, but not to the longer waves which we call heat.

It is evident that changes in the amount of the last two

267

gases will affect temperatures and climates on the earth; and theories have been worked out to account for ice ages by this means. Ozone is supposed by some to have a similar effect in the upper air. The screening effect of volcanic dust in the air also has been considered important in lowering temperatures.

Effects of Carbon Dioxide

Variations in the percentage of carbon dioxide have received much attention from geologists as a cause of mild climates or of ice ages, since it has been found by physicists that this gas, being transparent to light but opaque to heat, may serve to retain the earth's warmth somewhat like the glass covering of a greenhouse. Arrhenius has argued that a large increase in the amount of carbon dioxide in the atmosphere would probably give a mild climate to the poles, while an important decrease would probably bring on an ice age. If these statements are correct it remains for geologists to find out whether there are means by which the amount of this gas in the air can be sufficiently increased or decreased to produce these effects.

It may be set free in various ways, e.g., by the decay of organic matter, by the combustion of carbon compounds, by the heating of carbonates, such as limestones and dolomites, and as one of the gases coming from active volcanoes and certain springs. On the other hand, it may be removed from the atmosphere by the growth of green plants, which, with the aid of sunlight, use it up in their life processes and set free oxygen; or by the weathering of rocks resulting in the formation of limestone, etc.; or it may be dissolved by cold water, in which it is very soluble. By the warming of the water this gas would be given off again. Calcium carbonate (calcite or limestone) is soluble in water containing

CO_2, the bicarbonate being formed. If animals or plants remove the lime to form shells, etc., the extra amount of CO_2 is set free.

It is evident, then, that there are several ways in which carbon dioxide can be given off into the air or removed from it.

It has been supposed that a period of great volcanic activity would imply the giving off of an unusual supply of carbon dioxide to the air, and so would produce a universal mild climate. On the other hand, the growth of luxuriant vegetation and the storage of the plant remains as coal would remove carbon dioxide from the air, and, if the process went far enough, give rise to an ice age. It has been suggested that this applies to the Permo-carboniferous ice age, following the rich plant life of the coal measures. Chamberlin and other geologists have considered the process of rock weathering and the formation of limestone as more important. At a time of elevation of the land after diastrophic movements an unusually large surface of rock would be exposed to the action of CO_2, which would ultimately be locked up in the formation of limestone and dolomite, thus initiating glaciation. At a time of peneplaination when the sea encroached widely upon the continents, much less rock would be exposed and the processes which set free the gas would gain the upper hand. The air would be enriched in CO_2 and the climates would become mild.

It will be understood, of course, that the bald statements above are only suggestions of the line of thought, which would need to be greatly elaborated to give a true idea of the theories which have been proposed in which carbon dioxide plays a part. An excellent account of the subject is given in Chamberlin and Salisbury's Geology.[3]

Changes in the amount of carbon dioxide in the air must

be slow, and the theory does not account for the rapid climatic variations demanded by interglacial periods.

Effects of Volcanic Dust

Humphreys has recently worked out in some detail a theory of glaciation as caused by the screening action of volcanic dust in the air.[4] In the case of the explosions of Krakatoa, Pelee and Katmai the finest particles of dust were flung high into the air and reached the stratosphere, or isothermal region, perhaps even 25 to 50 miles above the surface, where there is no washing effect of snow or rain, and where they required from one to three years to return to the earth by the action of gravity. The finest dust particles "shut out solar radiation manifold more effectively than they hold back terrestrial radiation." The effect of the amount of dust sent out by Katmai, if long continued, would be to lower the temperature by several degrees Centigrade. "This small amount of solid material distributed once a year, or even once in two years, through the upper atmosphere, would be more than sufficient to maintain continuously, or nearly so, the low temperature requisite to the production of an ice age. . . . This quantity of dust yearly, during a period of 100,000 years, would produce a layer over the earth only about half a millimeter, or one-fiftieth of an inch thick, and therefore one could hardly expect to find any marked accumulation of it, even if it had once filled the atmosphere for much longer periods." [4]

He believes that within the last 160 years such violent volcanic explosions have lowered the average temperature as much as half a degree Centigrade, and that if the eruptions had been three or four times as numerous the snowline would have been depressed 300 meters (about 1,000 ft.), thus beginning a moderate ice age.

Not all volcanic eruptions give off dust, since the outflows of basic lava (basalt) are usually without important explosions. It is only in the case of rhyolitic, or acid, lavas, which are much less common than basic ones, that great quantities of dust are projected into the air.

As Humphreys suggests, it is for geologists to show whether great ice ages, such as that of the Permo-carboniferous and the Pleistocene, were accompanied by great and long-continued explosive eruptions; but thus far evidence of that nature is not forthcoming. It should be remembered, however, as remarked by Humphreys, that these volcanic explosions may have taken place in any part of the world with the same results, so that to disprove the theory finally it will be necessary to demonstrate that such eruptions did not occur in inaccessible regions, parts of the world still unexamined geologically, evidently a difficult thing to be certain of.

It is perhaps worthy of note that some early speculations suggested that great and long-continued volcanic eruptions were the cause of universal mild climates instead of cold ones, because of the carbon dioxide given off to the air by volcanoes. It might be of interest to inquire whether the two processes do not partially neutralize one another.

Though the theory is in some ways an attractive one, it seems to the present writer that most geologists will hesitate to assume such a steady and long-continued series of explosive eruptions as would be required for the hundreds of thousands of years of glaciation in the late Palæozoic and Pleistocene ice ages. Nevertheless vulcanism supplies a subordinate cause of refrigeration which might for a few years reinforce some more general cause in initiating a great snow sheet; which would be more or less self-perpetuating in later times.

Croll's Theory

Our ideas of the earth's early history are so closely related to various astronomic speculations that a number of writers have attempted to explain ice ages by reference to theories or facts of astronomy. The bearing of the nebular and planetesimal theories on the problems of ice ages has already been referred to in the discussion of Manson's attempt to account for periods of mild and of glacial climates by the earth's internal heat and the supposedly cloudy conditions of its early times of warmth.

A variation in the position of the earth's axis at different times has been appealed to as accounting for ancient climates; and changes in the shape of the earth's orbit have been assumed to provide a cause for ice ages. Changes in the amount of heat radiated from the sun directly affect the temperatures of the earth and a deficiency as compared with the present might well cause a glacial period, while an excess of radiation over the present might have given rise to the usually mild climates of the past. A variation in the temperature of space has been suggested also.

The astronomic theory which has been most carefully elaborated is that of a change of excentricity of the earth's orbit. This has been supported by Croll in two books which have attracted much attention from geologists;[5] and a modified form of the theory has been presented by Ball.

The orbit of the earth is not a circle but an ellipse with the sun at one of its foci. The elongation of the ellipse varies within certain limits. When the orbit is nearly circular the length of summer and winter is nearly equal and the distance of the earth from the sun is not very different in the two seasons. When the orbit is very elliptic the year is not equally divided. At present the northern hemisphere has about eight days more summer than winter and the southern hemisphere the reverse of this. Tables have been worked out to show the variations in the excentricity of the earth's orbit for three million years in the past, and Croll assumes that times of great excentricity imply ice ages. He has elaborated the theory in great detail, but only an outline of it need be given here.

When the northern hemisphere has its winter at a time of great excentricity the season will be thirty-six days longer and much colder than usual, and Croll assumes that in the northern parts snow will accumulate on the land to such a thickness that the short summer, though hot, will not remove the whole of it, and perpetual snow will be piled up producing an ice sheet. This will cause a strengthening of the northeasterly trade winds because of the greater difference in temperature between the cold arctic regions and the equator.

The northeastern trades will encroach therefore on the southern hemisphere and much of the warm water which now goes into the Gulf Stream will be deflected south along the Brazilian coast, thus warming the south Atlantic and further cooling the north Atlantic and the lands about it.

Owing to the precession of the equinoxes, in about 13,000 years the relation of summer and winter in the two hemispheres will be reversed, the northern summer will be long and moderate in temperature and the winter short. The glaciation will have disappeared, the northeastern trades will have been weakened and the southeastern strengthened, forcing a larger portion of the warm equatorial water along the north side of South America into the Gulf Stream, thus aiding in the warming up of the northern lands.

It will be seen that the process, as imagined by Croll, creates an alternation of cold and warm climates around the north Atlantic, i.e., of glacial and interglacial periods, the interglacial times being milder than the present climate and allowing temperate forests to thrive in Greenland and other arctic regions.

When the northern hemisphere is frigid the southern is warm and *vice versa;* since an interglacial period in the north means glaciation in the south.

As the total amount of heat reaching the earth from the sun is practically the same whether the orbit is nearly circular or very elliptic, it is evident, as Croll admitted, that excentricity in itself will not cause glaciation; but he believed that the circulation of the air, and, as a consequence, of the ocean, is greatly shifted, producing the effects just mentioned.

The increase or diminution of the Gulf Stream is a most important part of the machinery which he supposes can regulate the climate, and a rearrangement of the land, e.g., the opening of a channel to the Pacific at Panama, might completely change the conditions.

The theory as developed by Croll was long a favorite with geologists, but gradually difficulties were discovered in applying it, and at present it receives little attention.

The theory demands a large number of glaciations to correspond to the repeated times of great excentricity; and Croll believed that these occurred in the Miocene and the Eocene, etc., though there was little evidence to prove this. He explained the lack of evidence by the imperfection of the record, since glacial deposits are mainly formed on the land, while the geological records are mainly preserved in the sediments of the sea bottom.

The weakest point in the theory is the requirement that the two hemispheres must have opposite conditions. When Europe and North America are glaciated the southern lands should be warm. The equatorial regions, except for a little shifting of the zone of greatest heat to the south or the north, should not be refrigerated.

Now it is proved with certainty that there was a fall of temperature over the whole earth during the Pleistocene glaciation. The glaciers on lofty tropical mountains reached 3,000 feet lower than now, and there is a close resemblance between the boulder clays of North America and of Patagonia that would hardly be possible if one was many thousands of years older than the other. Two great extensions of the ice with an interglacial time are found all along the Pacific Cordillera of North and South America from Canada to Patagonia, the older till more widely spread and much more weathered than the later one.

This relationship is seen in British Columbia, the western United States, Ecuador, Peru, Bolivia, Argentina and Chile; as well as on Ruwenzori and Kenia in Central Africa. It cannot be doubted that the times of cold were simultaneous in the two hemispheres and at the equator, and no theory that demands alternate glaciation, north and south, is admissible.

Although the astronomic theory developed by Croll is no longer considered important as providing a cause for ice ages, several of his subsidiary ideas are of value. The effect of possible changes of prevailing winds and of ocean currents is worthy of consideration; and the power of a surface of snow to radiate heat into space and to perpetuate glacial conditions, as suggested by Croll, is undoubtedly one of the factors in initiating glaciation.

Drayson's Theory

Another astronomic theory of the causation of ice ages, suggested by Drayson in various articles many years ago, is that the plane of the ecliptic slowly changes its obliquity, reaching a maximum of $35\frac{1}{2}°$ instead of $23\frac{1}{2}°$, as at present.[6] He believed that the times of greatest tilt implied glaciation, and that the maximum tilt corresponded to the maximum glaciation. Astronomers do not seem convinced of the accuracy of his conclusions; but A. H. Barley and R. A. Marriott strongly support his theory. An obliquity curve for the last 31,756 years, worked out by Sir Algernon F. R. de Horsay, indicates that the last ice age began 21,000 years ago, reached its climax 13,000 or 14,000 years ago, and ended about 6,000 years ago; and Barley and Marriott have collected a number of estimates of the end of the ice age as occurring about 6,000 or 7,000 years ago, in support of the correctness of the Draysonian theory.

From the evidence given in former chapters, showing that the ice began to retreat from the Mississippi Valley about 30,000 years ago and that Niagara was freed from ice about 25,000 years ago, it will be seen that the theory does not fit the known facts, and therefore cannot be correct. In any case it is not evident why an increase in obliquity of the axis should produce important glaciation, since the total amount of heat received by the earth from the sun would be unchanged.

The Drayson suggestion differs from Croll's in providing for glaciation in both hemispheres at once, instead of alternate glaciation at the opposite poles.

Shifting of the Earth's Axis

The puzzling features of the Permo-Carboniferous ice age, when regions now subtropical and tropical were ice covered, in former days led some geologists to assume a shifting of the earth's axis of rotation to the center of the Indian Ocean, so that the glaciated regions might be grouped around the south pole. There was something to be said for this assumption as long as the known glaciation was confined to India, South Africa and Australia; but the discovery of a great area of glaciation in South America, which would be within the tropics under that arrangement, robbed the suggested change of any value unless combined with a shifting of the continents along the lines proposed by Wegener.

It should be noted that a south pole in the middle of the Indian Ocean would mean a north pole near the border of Mexico with Texas; and that powerful glaciation should occur in that region. The nearest known outcrop of tillite is on a small scale in Oklahoma 400 miles to the north.

Changes in the Radiation of the Sun

By far the larger part of the energy available on the earth's surface comes from the sun in the form of radiant heat, light, electricity, etc., and it is evident that any change in the amount of this radiation must affect terrestrial climates. Usually it is the sun's heat that is thought of in this connection, and it is known that this varies in intensity from time to time. It is stated that these variations may amount to 4 or 5 per cent; and a lowering to this extent, if long continued, would mean a very serious change in temperature. These variations are in both directions and are of short duration so far as known.

There are, however, stars which vary greatly in brightness and any planets revolving about them must undergo corresponding changes in temperature. Is our sun a long term variable star and may our mild periods and ice ages be accounted for as resulting from its variations?

One variation, that of the number of spots visible on its disk, has long been known, and has been found to have a fairly regular period, averaging a little over eleven years in its cycle. Many attempts have been made to correlate this period with cycles of weather on the earth, but it cannot be said that anything very definite has thus far been determined.

So far as ice ages are concerned it is clear that no short cycle will account for them; but may there not be similar though far more extensive solar changes of much longer span that would be effective?

In a recent book on "Climatic Changes," by Huntington and Visher, a Solar Cyclonic Hypothesis has been elaborately worked out to account, not only for ice ages, but also for times of aridity; and the same subject has been more briefly discussed by Humphreys and others.

What appears to be a curious paradox results from their investigations. During sun spots, which are a sort of solar cyclone, the sun radiates more energy than when its surface is quiescent; but on the earth sun spot periods appear to coincide with cooler temperatures. The sun sends more heat than usual and the earth, as a result, is not warmer but cooler than usual!

Huntington and Visher account for this by showing that during sun spot periods there is greater storminess on the earth and the storm paths are farther north than usual. The stronger winds thus caused cool the surface of the earth somewhat and carry the extra heat to higher levels where it is lost, so that the earth actually loses heat in the process.

Humphreys gives another explanation of the matter. During sun spots more ultra violet rays than usual are emitted. These transform part of the oxygen of the isothermal region into ozone which has the same screening effect as the volcanic dust referred to on a former page.

Huntington and Visher show from historic evidence in Europe and Asia and from the annual rings of the Californian sequoias that the fourteenth century was a time of unusual cold and stress. This they connect with a large number of sunspots visible to the naked eye as recorded in Chinese literature.

If such a relationship of numerous sunspots and cooler temperatures on the earth can be proved within historic times, it may be supposed that more extreme solar changes would largely account for times of glaciation and times of mild climate in the past. The conclusion is reached that solar changes are the most important causes of ice ages and of arid periods in the past, though terrestrial changes, such as the enlargement or diminution of the land surface and the deflection of ocean currents, play a part also.

The book is very interesting and its main conclusion deserve careful consideration, but some of the statemen it contains need correction.

It is suggested that in the ice age the snowline was as depressed near the equator, as shown by a lowering of 1,500 feet in Venezuela.[7] In reality the depressio from 3,000 to 4,000 feet in Ecuador immediately un r the equator, and about the same in the equatorial m m s of Africa.

The statement that "geologists are almost u sally agreed that the lands were exceptionally extensiv a also high, especially in low latitudes," in the Permi s not correspond to the modern proofs that the ice sh eached sea level on all four of the glaciated contine s shown in former chapters.

The interesting suggestion is made that th e re deserts in high latitudes while low latitudes were i c red in the Permian. Is it not possible that the deser itions came later than the glaciation?

Although variations in the sun's radi io seem to provide a very simple explanation of ice ages and other changes in terrestrial climates, it should be remembered that we have no positive proof that variations of sufficient magnitude occur, and that the supposed causes of such variations are very indefinite and may never have come into effect. The irregular spacing of ice ages in the earth's history and their widely different severity do not fit into any long range rhythm of solar variation but rather suggest some more accidental outside cause.

Changes in Temperature of Space

The suggestion has been made that the temperature of space may vary in different parts and that ice ages may result when the solar system passes through an unusually cold portion, while mild conditions to the poles would be caused by portions of space warmer than usual.

The expression "temperature of space" is not a very good one, since theoretically space itself is nothing and cannot have a temperature. In reality what is meant is the temperature of a body in space, i.e., outside of the atmosphere of our earth, which has a blanketing effect on the earth's surface. The temperature of the moon or of a meteorite is what is meant.

Some physicists state that the temperature of space is the absolute zero,—491° Fahr., but it is certain that innumerable radiations of light and heat, and no doubt other varieties of motion, are traversing space in all possible directions. All the millions of stars are sending off such radiations, so that every planet and meteorite is perpetually bombarded from all sides, and their temperature, even in the absence of our sun, must be above the absolute zero. What the heating effect of these radiations would be I have not seen estimated.

The fact that sounding balloons show a nearly constant temperature of about — 90° Fahr. above 13 miles of elevation, where the extremely rare outer atmosphere can have very little blanketing effect, suggests that a body in space, exposed to the radiations of the sun and stars, would stand about at — 90° Fahr.

To what extent the temperature of a body belonging to our solar system would probably vary by reason of its motion into regions of more thickly or more thinly scattered stars has probably never been estimated, but it should be remembered that on our earth a depression of 5° or 10° Cent. would probably bring on an ice age, and a rise of a similar amount would probably provide for the known mild climate of northern and southern regions during most of the world's history.

Estimates of the amount of light radiated to us by the stars show only a very small percentage as compared with that coming from the sun. There must be, however, a great amount of dark radiation in addition to what is of visible wave lengths.

Great areas of dark nebulous matter are being discovered in the sky, perhaps even approaching the amount of bright matter visible, and it may be supposed that much of the dark matter is far above the absolute zero in temperature and hence must be radiating heat rays. If the bodies in space are cooling down one would expect to find them in all stages from the hottest to the coldest, and many of them may be just below red heat, and therefore invisible, though still several hundred degrees above the absolute zero.

It is possible, also, that dark diffused matter might have a screening effect at some times, diminishing the amount of radiation reaching the earth from the sun, as well as from the stars, and thus lowering our temperature.

It is suggested by Shapley that there are great diffuse clouds of nebulosity, some bright, most of them dark. The probability that stars moving in the general direction of such clouds will encounter this material is very high, for the clouds fill enormous volumes of space. Collision with a dense part of such a cloud would produce a nova; and our sun may have been a miniature nova or a friction variable.[8]

This would account for periods of universal mild climate on the earth, but not for periods of cold.

Such vague and accidental causes for climatic changes should be appealed to only as a last resort unless positive proof at some time becomes available showing that an event of the kind actually took place.

Conclusions

In the foregoing chapters an outline has been given of the principal theories which have been put forward to account for ice ages; and it is evident that no single theory is universally accepted and that difficulties have been raised against all of them. It may be expected that the present writer, after pointing out defects in all the previous attempts to solve the tangled problems of glacial periods, should propose something which he considers more satisfactory. This I do not feel competent to undertake. During many years of study of glaciation I have hoped to find a solution of the difficulties in several theories at different times but have always encountered some point where they failed.

One must account for a world-wide refrigeration affecting all zones of both hemispheres at the same time, but with a focussing of great ice sheets in special regions, as near the north Atlantic in the Pleistocene; and one must provide for the relatively rapid alternations of cold and warm conditions demanded by a succession of glacial and interglacial periods.

In my opinion no single cause, as expressed in one of the theories proposed, can accomplish this; and any final solution of the complicated problems involved must come from a conjunction of general and local causes. Some combination of astronomic, geologic and atmospheric conditions seems to be necessary to produce such catastrophic events in the world's history. The rarity of such a conjunction would account for the comparatively few and irregularly spaced times of glaciation which interrupt the usual monotonous continuity of mild conditions shown by palæontology.

REFERENCES

1. HARMER, *Geol. Soc. Lon.*, 1901, Vol. 57, pp. 405-478.
2. BROOKS, *Quar. Jour. Roy. Meteor. Soc.*, Vol. 40, 1914, pp. 53-70; also Vol. 47.
3. Vol. II, pp. 658, etc.; also Vol. III, pp. 432, etc.
4. *Physics of the Air*, 1920, pp. 569, etc.
5. *Climate and Time*, 1875; and *Climate and Cosmology*, 1886.
6. *The Drayson Problem*, A. H. BARLEY; and *Warmer Winters and the Sun's Tilt, The Ice Age Fully Explained*, R. A. MARRIOTT, Exeter.
7. *Climatic Changes, their Nature and Causes*, HUNTINGTON and VISHER, 1922.
8. "A Possible Factor in Geological Climates," HARLOW SHAPLEY, *Jour. Geol.;* Vol. 29, 1921.
   The Earth, Its Origin, History and Physical Constitution, HAROLD JEFFREYS, Cambridge, 1924. This book was received after the final chapter of the foregoing work was complete for the printer. All the subjects are handled in a highly mathematical form and some of the conclusions reached are questionable because of unfamiliarity with the geological premises involved.. The discussion of the causes of ice ages is interesting but inconclusive.

INDEX OF AUTHORS

GENERAL INDEX

Abberley hill breccia, 93
Absolute zero, xv
Adelaide, 213
Afghanistan, 108
Africa, 239
 Dwyka in, 138
 Permo-Carboniferous glaciation in, 115, 138
Africa, Central, 42
 mesozoic glaciation in, 85
Africa, South, *see* South Africa
Age, Ice, *see* Ice Age
Age of reptiles, 84
Alaska, xxvii, xxxvii, 12, 86, 88, 191, 195, 196, 197
Alaskan boundary, xxvii, xxix, xxxv
Alaskan Permo-Carboniferous Tillite, 180
Alberta, 56, 84
Algonkian glaciation, 78
Alpine glaciers, xvi
Alps, 40, 157
 ice age in, 37
 pleistocene record of, 37
America, interglacial periods in, 19
America, North, *see* North America
America, South, *see* South America
Andes, 40, 42
Antarctica, xvii, 9, 64
 glaciation in, 43, 53, 82
Appalachian mountains, 202, 203
Appila Gorge, 214
Archæan glaciation, 78
Archæan peneplain, 229
Arctic regions, xvii
Area of cobalt series, 225
Argentina, 41, 165, 166, 167, 172
 glaciation in, 162
Arroya Negro, 163
Asia, Central, xvi
 early Cambrian glaciation in, 210
 ice age in, 38

Asia, Permo-Carboniferous deposits in, 108
 Pleistocene glaciation in, 38
Asiatic glaciation, character of, 109
 extent of, 109
Astronomical theories of glaciation, 246, 267
Atmospheric theories of glaciation, 267
Australia, 43, 211, 213, 239
 effect of glaciation on flora and fauna in, 150
 late Palæozoic glaciation in, 139, 149
 references to, 152
Australia, South, 143, 145, 213, 217
Australia, West, 229
Australian glacial beds, 142
Axis of earth, shifting of, 276

Baberton Rocks, 124
Bacchus Marsh, 139, 141, 144
Bahia Blanca, 163
Baltic ice sheets, 10
Banded clays, seasonal, xxxviii
Banding, seasonal, in Sudbury rocks, 233, 234
 in Timiskamian rocks, 234
Barakar, 112
Barriers, ice sheets combined with, 62
Beaches, marine, 66
 raised, 65
Bearer River, 196
Belgian Congo, 132
Belgium, 191
Blackman's Island, 178
Blaini conglomerate, 211
Bokkeveld, 195
Bolivia, 42, 170, 171, 173
 glacial conglomerates in, 169
Boston, 179
Boulder, Dalradian, 228
 from base of Cambrian, 217
Boulder beds, 112

289